The Last Imaginary Place

The Last Imaginary Place

A Human History of
the Arctic World

Robert McGhee

OXFORD
UNIVERSITY PRESS

2005

OXFORD

UNIVERSITY PRESS

Oxford University Press, Inc., publishes works that
further Oxford University's objective of excellence
in research, scholarship, and education.

Oxford   New York
Auckland   Cape Town   Dar es Salaam   Hong Kong   Karachi
Kuala Lumpur   Madrid   Melbourne   Mexico City   Nairobi
New Delhi   Shanghai   Taipei   Toronto

With offices in
Argentina   Austria   Brazil   Chile   Czech Republic   France   Greece
Guatemala   Hungary   Italy   Japan   Poland   Portugal   Singapore
South Korea   Switzerland   Thailand   Turkey   Ukraine   Vietnam

First published by Key Porter Books, Canada, 2005

Published by Oxford University Press, Inc.
198 Madison Avenue, New York, New York, 10016

www.oup.com

Oxford is a registered trademark of Oxford University Press

Library of Congress Cataloging-in-Publication Data

McGhee, Robert.
The last imaginary place : a human history of the Artic world / Robert
McGhee.
p. cm.
Includes bibliographic references and index.
ISBN-13: 978-0-19-518368-9
ISBN-10: 0-19-518368-1
Arctic regions—History. I. Title.

G606.M39 2004
909'.0913—dc22

2004066296

Printing number: 9 8 7 6 5 4 3 2 1

Printed in the United States of America
on acid-free paper

Contents

ACKNOWLEDGEMENTS

It is impossible to acknowledge the debts that I owe to so many people who stimulated my fascination for the Arctic, who made it possible for me to travel and work in the region, and who taught me what I know of it. Farley Mowat's early books on the Inuit of the Barren Grounds kindled my first interest, and his writing over the subsequent years continues to broaden my appreciation for his perception and humanity. The late William E. Taylor, Jr. introduced me to Arctic archaeology, and encouraged me to make my living at it. This was possible only through the help, patience and good humour of the dozens of students and colleagues who worked with me in field camps over the decades. Of these I must make individual mention of Dale Russell, whose company I enjoyed over several summers from the Mackenzie Delta to the High Arctic, and Jim Tuck, who introduced me to the archaeology of Newfoundland and Labrador and who mentored me in the Elizabethan archaeology of Baffin Island. Louwrens Hacquebord made it possible for Jim and me to visit Svalbard, and to experience this very different part of the Arctic. Hansjurgen Müller-Beck introduced me to the European community of Arctic research, and to investigations in Siberia. The time that I spent in Chukotka would not have been possible without the continuous assistance of Mikhail Bronshtein, and I wish to express my particular gratitude to Misha and his wife Lena for their hospitality and generosity. I am also grateful to the people of Qausuittuq who have helped me over the years that I have worked near their community.

I wish to thank Anna Porter of Key Porter Books for initially proposing this project, and Michael Mouland for coordinating its production. Earlier versions of the text were read by Hugh Brody, Igor Krupnik, Peter Jull, David Morrison and Pat Sutherland, and I appreciate their helpful comments. I am grateful to all who provided photographs. I also wish to thank the librarians at the Canadian Museum of Civilization, who cheerfully and efficiently tracked down the often obscure publications that I requested.

Finally, I wish to express my gratitude to Pat Sutherland for the support that she provided to me in the completion of this project as in so many other areas of life.

P RELUDE :

AN ARCTIC VISION

I FIRST BECAME AWARE OF THE ARCTIC during the 1950s, when I was a teenager whose world was limited to the well-tended farmlands and tame urban environments of Ontario. It is hard to remember how exotic most remote places seemed to us in those days, when air travel was a rare luxury and television offered only a few blurry channels depicting life in New York, London or occasionally Toronto. The little we knew of distant environments came mostly from books and movies, and from the imaginary journeys that they stimulated. Our ignorance gave travel writers the freedom to create visions of romantic landscapes and exotic peoples that at times were only tenuously related to reality. Like many Canadians of the 1950s, I was stirred by Farley Mowat's books on the desperate plight of the Barren Grounds Inuit, telling of a hunting people who had stepped directly from the Stone Age into a world of intrepid bush-pilots, hard-drinking fur traders and heartless bureaucrats. To Mowat's readers the remote landscapes of the Arctic were enhanced by the suggestion that they served as a meeting-place between a dramatically alien present and an ancient past.

Soon my imagination was awash with Arctic images that had flowed from books: a wild land of brilliant light, keen winds, pure ice and endless panoramas. Broad rivers of migrating caribou flowed northwards under the midnight sun across limitless plains carpeted with dwarf willows and tiny wildflowers. Bands of fur-clothed hunters waited patiently at river crossings where the ancestors of people and caribou had met every year for uncounted generations. In the polar night they travelled by dogsled on icefields where they encountered great white bears, or hunted walrus among the ice-floes that had doomed the expeditions of nineteenth-century explorers. Across the pole was another Arctic, a shadowy land peopled by Stalin's victims and known from rumour and propaganda. Superimposed on both Arctics was the science-fiction world of electronic warfare, still under construction, a ring of secret and isolated radar domes and sealed metal shelters in which technicians lived as they would in space-stations on a distant planet.

The Arctic of my young imagination is immediately recognizable in Sherrill Grace's perceptive *Canada and the Idea of North*. She describes the presentation of the northern world in mid-twentieth-century novels and documentary films as "a place of masculine romance," its wonders depicted in narratives that "construct the North… as a space for virile, white male adventure in a harsh but magnificent, unspoiled landscape waiting to be discovered, charted, painted, and photographed *as if for the first time*." This Arctic is a land that irresistibly attracts "those special few who can go North… and return safely to tell the tale."

Across the pole, my Russian contemporaries were experiencing the same pull towards the vast magnetic spaces of Siberia. In his study *Between Heaven and Hell: The Myth of Siberia in Russian Culture*, Yuri Slezkine describes how Soviet literature of the period was populated by "dozens of orphaned fictional youngsters [who] stampeded out of the soulless old capitals to a land where the snow never melted and where men kept their word."

This was the Arctic that I longed to experience, but it was an ambition that was not easily achieved at the beginning of the 1960s. In 1965 archaeology finally took me north as field assistant on a crew investigating one of the earliest indications then known of human occupation in Arctic Canada. Thirty-five years have passed, yet that first experience of the Arctic remains vivid: the long droning flight northwards from Yellowknife, past treeline and over the Barren Grounds pocked with ice-covered lakes; the clatter of gravel and billows of dust as the propeller-driven DC-4 ploughed along the dirt airstrip beside the DEW-line radar station; the chill air that filled the cabin as we pulled on parkas and caps to climb down into the brilliant June night. Outside the low sunlight flooded colour over snowdrifts, small and smiling people, old women with tattooed faces, howling dogs, small houses of unpainted plywood, and an icy stream flowing through a natural rock-garden of tiny yellow poppies and purple saxifrage. The Arctic of my fantasy easily survived this first encounter with reality, and I was immediately enchanted.

In his illuminating *Landscape and Memory*, Simon Schama explores the intimate links that exist between terrain, mythology and culture, and persuasively argues that these links condition our reactions to the environments in which we find ourselves. In Schama's view, the scientific description of a location, such as might be produced by a methodical geographer, is only the bare beginning of a definition of place. "Landscape" also comprises the mythology, history, artistic conventions and all the other cultural factors and associations that allow us to define a homeland, a wilderness, a terrain of scenic marvels worthy of sightseeing, or a land so dangerous that it inspires dread and avoidance. I believe that the concept of "Arctic" that is held in the minds and imaginations of those of us who live in more temperate parts of the world owes far more to these associations than it does to the actual physical reality of the North.

Most of the books that set out to portray the Arctic have ignored this component of our multi-layered human reality. They describe the temperatures and the seasons of the lands around the pole, the many forms of sea ice and their vital importance to the life of the region, the marvellous animals and tiny resilient plants, the ingenious, resourceful natives, the intrepid explorers and investigators who brought the territory into the realm of science. Many of these books are thorough, accurate and valuable sources of information on the Arctic regions; I make no claim that the present venture will be such a comprehensive source. In the following chapters the sea ice and the midnight sun, the whales and walrus and reindeer, the flaring aurora and the endless winter night are viewed only as scenes and players in the human history of the polar zone. The book attempts to present this history not as something that occurred to bizarre and alien peoples isolated from the rest of the world, but as a part of the global history of human endeavour. By understanding that the Arctic is peopled with human communities, with human ambitions and concerns that are recognizably related to those that we know from more familiar regions, the Arctic itself becomes more familiar, more comfortable.

As part of this process, we also confront the relationship between Arctic experience and the reports and rumours of that experience that have reached more southerly regions. These stories, the true and the false, have gradually accumulated to form the vision of a distant and fantastic Arctic as seen through the window of Western culture. This image of the Arctic as a world apart, where the laws of science and society may be in abeyance, is informed not only by what we hear from present-day observers or see when we visit the area ourselves. It is also moulded by a view of the Arctic that comes down to us from the distant past, when the region was as alien and as impossible for most people to reach as another planet. For millennia this Arctic vision has successfully absorbed the hearsay evidence of travellers' tales, the accelerating flow of scientific information, and in recent years even the tedium of government reports, while retaining its aura of wonder. This Arctic is not so much a region as a dream: the dream of a unique, unattainable and compellingly attractive world. It is the last imaginary place.

1 AFTER THE ICE AGE

I ONCE SPENT A FEW HOURS IN THE ICE AGE. It was a brilliant July day, the sun's heat comfortably tempered by a cool wind sweeping down from the frozen ocean beyond the ranges to the north. We were sitting in the sunny mouth of a small cave, at the base of a limestone outcrop that protruded like an eroded molar from the hills of the northern Yukon. My friend Jacques Cinq-Mars had discovered Bluefish Cave, and had spent several summers here carefully excavating the bones of ancient animals and the preserved traces of early human activities. Listening to him talk about the place, I idly surveyed the view to the valley below us, and the distant Old Crow Flats sprinkled with shining lakes and veined with channels of running water. It was the same landscape that had been scanned from this lookout by hunters who lived here 20,000 and more years ago, when the view included wandering herds of mammoths, horses and muskoxen as well as the caribou that can still be seen surging over these flats during their annual migrations. The past and present slide together at Bluefish Cave, a place where so little has changed that it seems the ancient past might be glimpsed out of the corner of your eye.

The ice age that gripped the earth some tens of thousands of years ago holds a continuing fascination because it was an alien world, and

11

yet it existed where many of us live today. It is easier to imagine ice age conditions when one is at Bluefish Cave than when one is in the valley of the Thames or the Rhine, or on the shores of Lake Ontario, yet we know that these southern places once had Arctic-like environments resembling that of the northern Yukon today. These environments are part of the history of most human populations; they were the homelands of our ancestors at a time when they were developing most of the characteristics that we think of as truly human. The allure of the Arctic may be linked to this fascination with the Ice Age. Perhaps it is not too implausible to suggest that we see in the Arctic a lost but somehow still familiar ancient world that we once knew well, but from which we have become separated by hundreds of generations of trial and effort in a world of constant change and increasing complexity.

The fact that we think of the "Ice Age" as an established fact of prehistory is one of the triumphs of nineteenth-century science. For most of that century, science and theology fought an extended battle over the nature of the world and of mankind. Traditional and religious thinkers believed that the earth had been created a few thousand years ago, and that aside from a few divinely induced cataclysms and the improvements wrought by humankind, it was much as it had always been. Geology was the science that disturbed that settled view of the world. Geologists were convinced that the earth was an inconceivably ancient planet that had developed through aeons of slow and continuous change, and that its cargo of creatures had gradually evolved from simple life forms to the vast variety of the present day.

For public demonstrations of this theory, nothing could be more appropriate than the concept of the Ice Age. Scientists could point to clear geological evidence of such a time: ice-scoured rock and boulder debris left by melting ice indicated that the glaciers which now capped the European Alps, the mountain massifs of Central Asia and the mountain chains of western North America had once advanced over the surrounding countrysides. From Scandinavia and northern Canada immense continental ice-sheets had spread southwards to bury much of Europe and North America beneath a kilometre or more of ice. To the south of the ice-sheets, cold environments extended deep into

Glaciers similar to those that still cover portions of High Arctic islands once extended across much of Europe and the northern half of North America. (Patricia Sutherland)

what was now the temperate zone, supporting animal species currently found only in the polar regions. Palaeontology and archaeology, the newly developed daughter-sciences of geology, challenged the religious creation story by describing an Ice Age Europe in which reindeer, muskoxen and wild horses were hunted by cave-dwelling humans in the shadow of looming glaciers. More controversially, they assembled Ice Age fossils to reconstruct the skeletons of animals that no longer existed: mammoths and mastodons, sabre-toothed cats and giant cave-bears. And having successfully argued that humans existed alongside these bizarre animals, they went on to discover the bones of ancestral human-like creatures that were beyond the range of forms that could have been created in the image of the Christian God. In the nineteenth-century imagination the Ice Age became the paradigm for a time when the human race was in its infancy, and shared an extraordinary Arctic world with a menagerie of beasts stranger than any discovered by contemporary explorers.

We now know that the concept of an "Ice Age" is a flagrant oversimplification of a long and complex series of events that took place over the course of a billion years. The reasons for the earth's periodic cooling, and the advance and retreat of glaciers, are not understood although several possible causes have been suggested: minor changes in the earth's orbit, alterations in the amount of carbon dioxide in the atmosphere, or the slow movements of continents and consequent effects on the circulation of oceanic or atmospheric currents. For whatever reasons, episodes of worldwide cooling have occurred throughout much of the planet's history, and have been much more common in some periods than in others. The current series of episodes began about three million years ago, and has comprised at least twenty major glacial events.

Some have suggested that the ice ages of the past three million years played a crucial role in the development of human life. Changes in temperature and rainfall patterns during ice ages resulted in the spread of savannah grassland conditions throughout much of Africa, at the expense of tropical forests. These conditions may have encouraged the bipedal posture adopted by our ape-like ancestors, freeing their forelimbs for other uses and setting off the evolutionary chain of events that produced a creature whose hands were capable of carrying and manipulating objects, and whose anatomical potential included a large braincase and a voice capable of complex communication.

By about a million years ago our ancestors had spread out of Africa and into the temperate regions of the Old World, where they were subject to much more extreme environmental changes when the climate periodically cooled to ice-age conditions. These creatures were now recognizably human, of an archaic small-brained form known to palaeontologists as *Homo erectus*. It is doubtful that they had the skills and technology to survive full glacial conditions, but in both Europe and China there is evidence that some groups had the use of fire and had adapted to relatively cold climates.

Fully modern humans appeared during the last ice age (the one we may give a historical title as the Ice Age), at some time after about 50,000 years ago. Whether they came out of Africa or developed elsewhere from earlier archaic populations is still argued avidly by scientists. In either case, these new creatures with large brains and an aptitude for inventing and using tools quickly spread around the temperate regions of the world. As the Ice Age reached its height, between about 30,000 and 15,000 years ago, they adapted to it by developing throwing-spears and fish-hooks to obtain food, fire and sewn clothing for warmth, bark torches for light and artists' crayons to portray their world and their lives on the walls of caves.

The treeless tundra and steppe environments that invaded earth's mid-latitudes during the last ice age supported vast herds of grazing animals. These were ideal conditions for hunters who had developed the weapons and skills to undertake cooperative hunts for caribou, horses, wild oxen and other animals as large as mammoths. The major leap from slowly developing archaic forms to modern humans, with their rapidly evolving cultures and technologies, occurred at a time when ice-age environmental conditions prevailed in most temperate regions of the world. It has been argued that the challenges imposed by these Arctic-like conditions were an important stimulus to the developing ingenuity of modern humans. A century ago, when this theory first appeared, it seemed to support the new racial politics that claimed natural superiority for the European peoples, Friedrich Nietzsche's new Hyperboreans whose northern ancestors had been honed and tested against cold and ice. These ideas seemed to accord with the remarkable archaeological discoveries that were then being made, revealing the sophisticated hunting technologies and artistic accomplishment of the people who lived in European caves at the time of the last ice age. Since then it has become clear that similar levels of accomplishment characterized all human groups of the period, not just those whose remains were well preserved in the cold and dry limestone caverns of Europe. However, the vast herds of animals grazing the mid-latitude grasslands and tundras across Eurasia and America in the last ice age may have been the economic base that allowed some hunting peoples to develop the cultural and social complexity that served as the springboard to later human accomplishments.

Of course, the peoples of ice age Europe, Asia and America did not have to contend with the most fundamental characteristic of polar Arctic regions: the long night of winter when cold and darkness make hunting almost impossible, and when most animals have disappeared through migration to the south or to refuge beneath the sea ice. The usual assumption has been that humans developed the skills enabling them to live under such conditions only after the Ice Age had ended. Recently, however, Russian archaeologist Vladimir Pitul'ko reported the discovery of a small collection of tools associated with radiocarbon dates of between 25,000 and 30,000 years ago, on the Yana River near the coast of Siberia at a latitude of greater than 70° north. At about the same time, and at a slightly lower latitude, early hunters may have been crossing the land bridge that joined Siberia to Alaska during the Ice Age, to become the early ancestors of native American peoples.

When the Ice Age ended, about 11,000 years ago, it was as though a giant climatic switch had been thrown. We used to think that major changes in the earth's climate happen gradually, over centuries or millennia, but we now suspect that they can and do occur in periods of a decade or less, and perhaps at times as abruptly as in a single year or a single season. For hunters in the river valleys of Europe, on Asiatic plains or across North America, an exceptionally hot summer or a warm winter followed by a prolonged change in the climate could mean that animal herds were not where they had always been before; that rivers once easily forded had become impassable torrents of meltwater; or that ice failed to form on lakes where it had always provided a platform for winter fishing and travel. A hunting people's life depends on an intricate knowledge of the animals they hunt and the environment in which they live, and climatic change would mean that their knowledge, accumulated over dozens or hundreds of generations, was suddenly obsolete.

Those families and bands that survived the first few years of the postglacial found themselves in a new world of constant change. While the climate is capable of altering virtually overnight, its effects on other elements of the environment proceed at a slower pace. The melting of continental glaciers took a few millennia to complete, and during that time forests and grasslands moved northwards to replace glacier-edge tundra as the biological zones of the northern hemisphere became established in something like their current form. The mid-latitude tundras that were home to Ice Age hunters expanded northwards while their southern boundaries were invaded by shrubby conifers, the pioneers of dense boreal forests that would themselves be replaced by deciduous forests, parklands, grasslands and deserts. Melting glaciers drained southwards to form huge icy lakes, or into mid-latitude inlets of the sea where walrus and whales swam among calving icebergs. As the earth's crust rose from beneath its burden of glacial ice these lakes and inlets drained, at times with terrifying suddenness as new channels opened.

By about 8,000 years ago the earth had been transformed into a semblance of its present state. Climates in most regions were significantly warmer than at present, and most human groups had lost all contact with the icy world that had been known to their ancestors, the world of snow, sea ice, walrus, reindeer, and the cold that sucked the life from humans unprotected by fire, shelter and heavy clothing. The hunting way of life became increasingly demanding for most peoples of the new postglacial age. Although the temperate and subtropical forests that advanced into their old homelands were biologically rich, the great variety of animals they supported were dispersed across small and specialized niches. Making a living from hunting such animals was a far more difficult enterprise than following the great herds of reindeer, muskoxen, horses and elephants that had roamed the tundra and northern grasslands of the Ice Age. Some bands of hunters, fortunate in their local circumstances or prepared to fight to maintain their ancestral livelihood, moved northwards with the animals and the open treeless environments on which they depended. For these groups the past 10,000 years saw a constant succession of adjustments, inventions, strife with changing neighbours, and adaptations to new home territories as they evolved into the hunting and herding peoples of the Arctic world.

Most human groups followed a different course. Bands became dispersed across the changing environments, each concentrating on the particular resources of their new and limited homelands. Certain bands began to think of themselves as river fishers, others as coastal shellfish collectors, still others as forest hunters who snared deer and smaller game. The new environments also provided an array of food resources that had been practically unknown to northern hunters:

seeds, nuts and roots often became the staple of their survival. Within a few millennia of the Ice Age, peoples living in southeastern Asian forests, the river valleys of western Asia and the highland plateaus of Mexico had established a livelihood on the seeds of wild plants. Particularly useful forms of plants were discovered and protected, and their favoured forms encouraged or replanted. This activity began the process of genetic modification that resulted in ancestral varieties of rice, barley, wheat and maize. Other regions saw practical experiments with tropical root-crops—yams, taro, manioc—and elsewhere with legumes, squashes and fruits. By about 8,000 years ago, while the last remnants of the continental ice sheets were still melting in northern Canada and Scandinavia, people from Japan to Mesopotamia to Central America were living in small agricultural villages. This was the first step in what seems to have been an inevitable chain of events leading to the establishment of ancient civilizations.

Many observers have characterized these developments as exemplifying the human spirit's triumphant progress. However, the development of civilization might more convincingly be described as a treadmill on which human groups found themselves toiling whenever they occupied an environment that was rich enough to support a large population. At first, plant foods provided a stable and secure resource in return for relatively little labour, and the new farming way of life offered an easier existence without the constant travel and the discomforts of temporary camps that had been a part of their ancestors' lives for so long. More children survived in farming villages, but as populations began to grow rapidly it became apparent that people would have to work harder than ever to avoid starvation. There were two ways to cope with the problem: they could devote themselves to more intensive farming, which involved an ever-increasing burden of labour, drudgery and stress on the local environment, or they could opt for the path of warfare, mounting attacks on neighbouring groups to gain access to their land, their stored food or the labour that they could provide as slaves or tribute-paying subjects. Most agricultural peoples followed some combination of these alternatives, with the result that by about 5,000 years ago the stage was set for the revolving cycle of empire-building, conquest and destruction that has affected most humans living between the tropics and the temperate zones to the present day. Only during the past century has this process begun to have a significant effect on the lives of those peoples whose ancient ancestors chose to remain hunters, and to follow the hunting environments of the Ice Age northwards into the lands that were to be known as the Arctic.

To urban peoples the known world has generally been seen as a series of concentric zones around the hub of the home city. With each step outward from the centre the world seems less civilized, the people stranger and less predictable, and the world itself more fantastic. The first cities developed in the temperate and subtropical zones, and their inhabitants saw the north and south as barbarian lands where one might expect to find fierce warriors, people who didn't understand the benefits of civilization but did appreciate commerce, and had access to rich sources of amber, ivory, furs, gold and precious stones. Beyond these zones lay even wilder lands that were barely habitable from the intensity of heat or cold. Here one might expect to find semi-human creatures, fantastic animals and regions where the land and the sea, even the sun and moon, behaved in ways that were beyond the normal laws or conditions of the world.

In records left by the scholars of early civilizations, the Arctic was one of those distant and fantastic lands. No traveller's tale of the North was too bizarre to be believed. It is startling to realize that remnants of this view still cling to the region. The Arctic is still a place that is seen primarily through the eyes of outsiders, a territory known to the world from explorers' narratives rather than from the writings, drawings and films of its own people. To most southerners the Arctic remains what it was to their counterparts centuries and perhaps even millennia ago: the ultimate otherworld.

2 A Distant Paradise:

The Arctic in Ancient

Thought

On a cold summer evening in 1959, I first heard the story of the mysterious tropical valley in the Arctic. The construction-camp bunkhouse in the British Columbia mountains had settled into its after-supper routine. Most of the younger guys were out working on their pickup trucks or had disappeared down the rutted logging road to the town at the bottom of the valley. The elders of the camp were lying in their bunks, smoking and listening to rain on the tarpaper roof, staring at the ceiling or leafing through tabloid newspapers, too weary after a wet day on a chainsaw to do more than talk and wait for sleep. Conversation drifted through the usual topics: memories of home in the Maritimes, Budapest or on the Saskatchewan farm; laments about the new government tax that was killing jobs in construction; war stories from Korea and Europe. An article in one of the tabloids asked whether Hitler was really dead or had escaped to some distant land, and somehow this led to talk of a tropical valley that was rumoured to be hidden far to the north, somewhere in Arctic Canada.

Most of the men had heard something of this secret valley. George the Finn thought that it probably existed because he had heard of a very similar place somewhere up in Lapland. Old Alex said that it was just an exaggeration of the hot springs up in the Nahanni country of

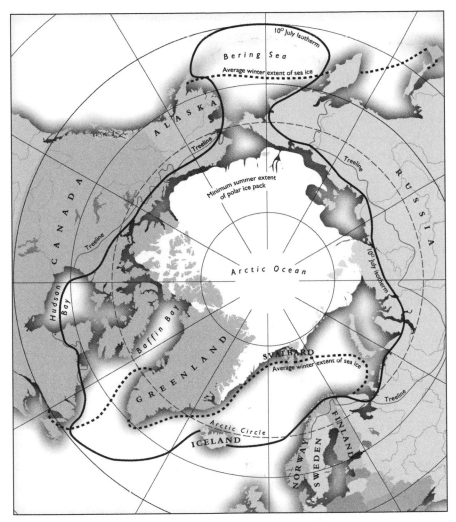

The Arctic Zone

the Northwest Territories. Others didn't believe it at all, but even the sceptics fell silent and listened as a bulldozer operator told us that his brother had met a bush pilot who had flown over the valley while lost in an Arctic storm. The pilot, he said, was amazed by the sight of palm forests in a wide valley surrounded by winter snows. The idea that somewhere to the north there might lie a hidden pocket with a climate that defied the laws of geography was eerily attractive, and none of us in the bunkhouse that night knew enough of the world to be certain that it was impossible. Nor did we know that the stories we were hearing on that rainy evening stretched back through European geographical lore for almost 3,000 years.

To the Greeks of the first millennium BC, the habitable world centred on the Mediterranean Sea. Here was found the temperate climate best suited to human life, subject neither to the freezing cold of the northern mountains nor the scorching heat of desert lands to the south. Greece was separated from central Europe by the broad and hostile Balkan mountains, which formed a formidable barrier to geographical knowledge. The word Arctic comes to our modern languages from that of the ancient Greeks. It derives from "Arktos," the Great Bear, the constellation that we call the Big Dipper, which circles the northern sky without setting. This celestial phenomenon was thought of as so uncanny that the land that lay beneath it might also be a place where the ordinary laws of nature and society would not hold.

Information about the north reached the Greeks by a route that passed to the east of the Balkan mountains, through the vast grasslands to the north and east of the Black Sea. These plains were inhabited by a nomadic people the Greeks called the Scythians; the most northerly people of the known world, their lands and those of allied tribes were said to be bounded on the north by the impenetrable "Rhipaean Mountains"—probably an amalgamation of rumours related to what are now known as the Carpathian, Caucasus and Altai ranges. From caves in these mountains issued Boreas, the bitter north wind that blighted northern regions and occasionally made life miserable even in the Greek archipelago.

To the north of the cold, dreary and hostile Rhipaean Mountains lay a fabled land where myth and geographical theory fused. This was an exotic and favoured Arctic world inhabited by the enviable "Hyperboreans"—the people who lived beyond Boreas. Protected by the mountains from the frigid blasts of the north wind, the Hyperboreans enjoyed an eternally pleasant climate in which there were no seasons and the trees bore fruit year-round. Their lives were passed in music, dance, serenity and comfort, untroubled by work, strife or disease. The Hyperboreans were immortals, but tiring of their easy lives after 1,000 years they decked themselves in garlands and leaped from a cliff into the sea.

The Hyperboreans and their Arctic paradise are a very ancient element of the Greeks' belief system, mentioned in the poetry of Hesiod as well as in a poem ascribed to Homer, both of which date to the beginning of written literature in the Greek language.

In parallel to their mythological knowledge of the north, the ancient Greeks began to accumulate factual information. The first step in Arctic exploration is usually ascribed to Pytheas, an astronomer and geographer from the colony of Massilia—now Marseilles. Only fragments of Pytheas's knowledge and adventures have survived the hazards of historical transmission, but they tell an intriguing story. About 330 BC he is said to have led an expedition northwards, perhaps to investigate the sources of tin and amber—two valuable commodities that were traded southwards to the Mediterranean. Beginning a journey that was to last for six years, the explorers sailed west through the Straits of Gibraltar and then northwards along the coastal trading route pioneered by earlier Phoenicean sailors. Pytheas reached Britain, where he visited the tin mines of Cornwall and explored other parts of the island; he even reported a circular temple (sometimes thought to be Stonehenge), which he described as the Hyperborean temple of Apollo, where the god was said to spend the winter months before returning to his shrine at Delphi in the spring. The explorer then continued northwards to investigate a land called Thule, where he had been told that the sun was visible at midnight at the time of the summer solstice.

Pytheas reported that after sailing northwards from Britain for six days, he arrived at Thule, whose inhabitants he described as barbarian farmers who had few domestic animals and little grain. Another day's sail to the north of Thule he encountered a fearsome barrier that he described as the "sea-lung," in which sea, earth and sky congealed into a single impassable element. The description is so vague that it might refer to a dense fog, or to loose sea ice undulating in the ocean swell, or perhaps to a combination of these phenomena which are so characteristic of Arctic seas. Whatever the sea-lung actually was, it marked the northern limit of Pytheas's explorations.

The location of Thule has puzzled geographers for more than 2,000 years. Iceland has been a perennial candidate, in view of its distance north of Britain, its high latitude where the "midnight sun" occurs at midsummer, and the proximity of sea-ice to its northern coast. However, Pytheas's description of the farmers who lived in the country of Thule is at odds with the fact that Iceland was almost certainly uninhabited at the time. It is more likely that he had sailed northeastwards, reaching Orkney or Shetland or even the western coast of Norway. Such a route would be more feasible for the small ships that Pytheas would have used, and his description of the Thuleens would fit the early Iron Age farming peoples of Scandinavia. Wherever Pytheas actually ended his northern journey, in subsequent centuries the name Thule or "Ultima Thule" became a vague designation for a country located at the farthest end of the northern world, which gradually acquired some of the paradisal elements of the Land of the Hyperboreans. The name floated about the northern portions of maps until the early twentieth century, when it finally acquired a fixed position: the Danish scholar Knut Rasmussen chose the name for a post that he established in northwestern Greenland, among the most northerly Inuit population in the world. The name was later transferred to the American airbase established nearby, which served as an important outpost in the Cold War of the late twentieth century. As we shall see, the name also acquired a more esoteric and malevolent connotation.

The mythological and geographical stories of the Classical world survived in ever more tenuous and distorted forms throughout the Dark Ages and the subsequent medieval period. In Europe, the northward spread of Christianity brought literacy and knowledge of a greater temporal as well as spiritual world, but also focused historical and geographical interest in biblical lands. The geographical worldview of medieval Europe is superbly encapsulated in the great world maps associated with the thirteenth-century English scholar Gervase of Tilbury. The most complex of these, a massive circular compilation over three metres in diameter assembled from thirty sheets of vellum, was preserved in the Monastery of Ebstorf in Germany until removed to Hanover where it was destroyed during a bombing raid in 1943. The plan of the world is superimposed on a figure of the crucified Christ, with his head to the east (top) of the

map and his feet in the western ocean. Jerusalem lies at his heart, at
the centre of the world. The Arctic regions are peripheral to this view
of the earth, and are represented as a northern extension of Asia
bounded on the north by an open sea containing a variety of islands.
The most notable feature is a range of mountains extending to the
Arctic coast, perhaps a successor of the Rhipaean Mountains of
antiquity. This range forms a barrier protecting Europe and southern
Asia from the Land of Gog and Magog, depicted as two figures feasting on human limbs, and probably representing the nomadic warriors
who occasionally erupted out of central Asia. The Land of the
Amazons, another myth of antiquity, has also been transposed to
Arctic Asia, marked by two large and well-armed women guarding a
land located just above the right hand of Christ.

Northern Europe is a relatively unadorned region of the Ebstorf
map, crossed by several unidentified rivers and centred on an area
containing drawings of an elk or deer, and a horned animal that may
be a wild ox; the similar Hereford map of the same period places a
bear in this region, perhaps representing the Russian forests. Off
Europe's western coast lies a string of islands including Anglia
(England), Scotia (Scotland), Hibernia (Ireland), Islandia (Iceland)
and Norwegia (Norway), forming a chain from south to north and
all barely separated from the mainland. The conventional form and
the lack of detail given to the northern European coasts and islands
suggest that although scholars recognized the existence of northern
lands as distant as Iceland and Norway, little was known of their
actual location or character. In a world that was seen as an imperfect
metaphor for the body of Christ, regions so distant from the heart
were of little interest or consequence.

The open and apparently navigable Arctic sea and the existence of
habitable northern islands are the only indications in the Ebstorf map
that the Arctic was still regarded as a region where human life could
flourish. However, the survival into medieval times of ancient concepts of Hyperborean wonders is hinted in the twelfth-century Book
of Leinster, a collection of writings including a mythic version of
Ireland's ancient history. Here it is reported that Caicher the Druid
led his people northwards from Scythia, and was driven into the

northern ocean by a storm "till at the end of a week they reached the great promontory which is northward from the Rhipaean Mountain, and in that promontory they found a spring with the taste of wine, and they feasted there, and were three days and three nights asleep there." Here Caicher prophesied that their descendants would travel unceasingly until they reached Ireland three hundred years from that day. To the author of the Book of Leinster, the Arctic was apparently still a place where marvels could be expected, and where one might sail through open seas from Scythia to Ireland.

The promise of an ice-free Arctic ocean was to tempt navigators of the following centuries, but early post-medieval maps also began to indicate a polar continent, adding another element that persists at the edge of current geographical awareness. The polar continent is best shown in the work of the Flemish geographer Gerhard Mercator, whose 1569 map served as a major source of information for Martin Frobisher and other early explorers of the Arctic. This map shows a roughly circular continent extending from a latitude of about 75° to 87° north, surrounding a polar sea centred on a single high mountain. The continent is divided into four islands by narrow channels that flow northwards towards the pole. George Best, a seafaring gentleman who wrote the most comprehensive account of Martin Frobisher's venture into Arctic Canada, provided a more detailed description of the Arctic as set out in this map:

> *For as Mercator mentioneth out of a probable Author, there was a Frier of Oxforde, a greate Mathematician, who himself went verye farre North, above 200 yeares agoe, and with an Astrolabe described almoste all of the lande aboute the Pole, finding it divided into foure partes or Ilandes, by foure greate guttes, indraftes, or channels, running violently, and delivering themselves into a monstrous receptacle, and swallowing sincke, with suche a violent force and currant, that a Shippe beyng entred never so little within one of these foure indraftes, cannot be holden backe by the force of any great winde, but runneth in headlong by that deepe swallowing sincke, into the bowels of the earth. Hee reporteth that the Southweast parte of that lande, is a fruitfull, and a holesome soyle. The Northeast part (in respect to England) is inhabited with a people called Pygmaie,*

whyche are not at the uttermoste above foure foote highe.... Al these indraftes & raging channels, runne directly towards a point under the Pole, where is also said to be a monstrous gret Mountain of wonderful gret height, & about 35 leagues in compasse at the foot.

The reference to the English mathematician with an astrolabe relates to a mysterious book titled *Inventio Fortunatæ,* which was said to have been presented to Edward III of England by a learned friar from Oxford who had travelled widely in the Arctic during the mid-1300s. The book has not been seen since the sixteenth century. It has been conjectured that the author was an English monk who visited the Norse settlements in Greenland during the mid-fourteenth century. The Arctic whirlpool, like the Pygmies noted by the author of the *Inventio* as occupants of the polar continents, seems to derive from a combination of Norse mythology and the geographical knowledge of Norse Greenlanders.

Mercator's map of the north was influenced by a second esoteric source, the narrative of the supposed fourteenth-century journeys of two Venetian brothers named Nicolo and Antonio Zeno. This narrative and an accompanying map had been published in 1558, purporting to tell of a series of explorations in the far north undertaken almost two centuries before. Several large and newly discovered northern lands were described: Estland, Frisland, Icaria, Estotiland, Drogco, as well as the relatively identifiable Islanda and Engronelant. These countries were said to be peopled with fantastic races of warriors, cannibals and Christians, and a picture was painted of an Arctic region that supported flourishing civilizations; the most northerly location described is the prosperous monastery of St. Thomas, associated with a region of hot springs at a latitude of almost 75° north on the eastern coast of Engronelant. The Zeno map was fraudulent, assembled from sources available to a mid-sixteenth-century Venetian author, but the fact that it could have been accepted by a geographer as knowledgeable as Mercator indicates that the concept of a mild and habitable Arctic was still alive in European consciousness.

This view was certainly held by John Dee, a friend of Mercator's

as well as astrologer and adviser to Elizabeth I of England, and
described as the most learned man in Europe. In 1580 Dee assembled information in support of an English claim to sovereignty over
the northern regions of North America, and stated that the British
King Arthur had

> ... not only Conquered Iseland, Groenland, and all the Northern Iles
> cumpassing unto Russia, But even unto the North Pole (in manner)
> did extend his Jurisdiction: And sent Colonies thither, and into the
> Isles between Scotland and Iseland. Whereby yt is probable that the
> late named Friseland Iseland is of the British ancient Discovery and
> possession: And also seeing Groeland beyond Groenland did receive
> their inhabitants by Arthur, yt is credible that the famous Iland
> Estotiland was by his folks posessed.

Dee's information apparently came from another book that has not
been seen since the sixteenth century, the *Gestæ Arthuri*, but may have
been embellished by his own imagined triumphs of British accomplishment in an Arctic that was favourable to human occupation.

George Best, in his 1578 account of the Frobisher voyages,
assembled detailed arguments to explain why the Arctic regions
must be habitable. He notes that temperature is not solely a function
of the angle at which the sun's rays strike the earth, but also of the
duration of the day, so that the endless day of Arctic summer allows
the development of temperate climatic conditions. During the winter, twilight and moonlight supply sufficient illumination, and
nature has provided the creatures of the country with fur, feathers
and other means of surviving the cold. This theory was also supported by the learned Renaissance cosmographer Guillaume Postel,
who stated that "... here under and aboute the Pole is beste habitation for man, and that they ever have continuall daye, and know not
what night or darkenesse meaneth."

The belief that the Arctic was a land of Hyperborean delights faded
only gradually as the explorers of the sixteenth to nineteenth centuries
brought back their hard-won geographical knowledge. In 1816 Mary
Shelley could still invoke the old dream when she used an Arctic expe-

dition as the framework for her story, *Frankenstein, or the Modern Prometheus*. In a letter to his sister written from St. Petersburg, her narrator, Robert Walton, describes the attractions of polar exploration:

> *I try in vain to be persuaded that the pole is the seat of frost and desolation; it ever presents itself to my imagination as the region of beauty and delight. There, Margaret, the sun is forever visible, its broad disk just skirting the horizon and diffusing a perpetual splendour. There— for with your leave, my sister, I will put some trust in preceding navigators—there snow and frost are banished; and, sailing over a calm sea, we may be wafted to a land surpassing in wonders and in beauty every region hitherto discovered on the habitable globe.... What may not be expected in a country of eternal light? I may there discover the wondrous power which attracts the needle and may regulate a thousand celestial observations that require only this voyage to render their seeming eccentricities consistent forever. I shall satiate my ardent curiosity with the sight of a part of the world never before visited, and may tread a land never before imprinted by the foot of man.*

Twenty years later, Edgar Allan Poe imagined a more sinister Antarctic version of a tropical polar region. In *A Narrative of Arthur Gordon Pym of Nantucket*, explorers force their way through the barrier of ice protecting the Antarctic sea. At high latitudes they find a calm and increasingly warm ocean, and a forested island that is home to a people resembling Polynesians. Escaping in a native canoe as the rest of the crew are massacred, the narrator is helplessly drawn southward by an accelerating current running towards the pole, where the story ends abruptly in a confusing maelstrom of heat and falling water. A few years earlier Poe had described an Antarctic polar whirlpool more vividly in *Manuscript Found in a Bottle*, which he later published with a concluding note that demonstrates the curious means by which culture transmits mythical information:

> *NOTE.— The "MS Found in a Bottle," was originally published in 1831* [1833, actually], *and it was not until many years afterwards that I became acquainted with the maps of Mercator, in which the ocean*

DISCOVERY OF THE NORTH POLE AND THE POLAR GULF SURROUNDING IT.

A large Ark riding at anchor; her crew starving, and nothing but a | JOHN B. SHELDEN, | I made this discovery on the night of the 25th of October, 1869.
drowned horse left for them to eat. | **DISCOVERER,** | The Pole is ten degrees high, and ten degrees wide, and round.
The Pole is a topaz or diamond. The water around the Pole never | MILLVILLE, N. J. | The outer wall is ice. The centre of the Gulf is ninety degrees north.
freezes.

The North Pole and surrounding open sea, as portrayed by John B. Shelden who claimed
to have sailed there in 1869. This hoax demonstrates the astonishing durability of
geographical concepts that were already ancient when portrayed by Mercator almost
three centuries earlier. (Anonymous, *Discovery of the North Pole and the Polar Gulf
Surrounding It,* 1870. Collection of Glenbow Museum, Calgary, Canada)

> *is represented as rushing, by four mouths, into the (northern) Polar
> Gulf, to be absorbed into the bowels of the earth; the Pole itself being
> represented by a black rock, towering to a prodigious height.*

Poe's vision of a polar entrance to an underworld may not have
come directly from Mercator, but both were likely drawn from the

same ultimate source. Poe must have been familiar with the views of an eccentric American army officer, John Symmes, who wrote and lectured widely during the early nineteenth century on the bizarre concept that broad entrances at the North and South poles gave access to the habitable interior of the earth. It seems likely that Symmes merely elaborated an idea that had been proposed by the seventeenth-century English astronomer Edmund Halley, the scholar who also discovered the periodic return of the comet that now bears his name. Halley had suggested that the earth was hollow and habitable, lit by luminous gases that occasionally escaped through polar orifices to cause the aurora. Barely a century separated Halley's work from the time when the Arctic geographical treatise known as *Inventio Fortunatæ* was known to exist. This obscure medieval document was treated as an authority by scholars of the early Renaissance, and may have provided the basis on which both Halley and Mercator constructed their views of the north polar zone.

Although the concept of polar marvels was eventually discarded by geographical scholarship, it survived as a literary and theoretical idea in the consciousness of European cultures. Jules Verne's nineteenth-century adventurers in *Journey to the Centre of the Earth* found the entrance to this fantastic region in an Arctic mountain, the Icelandic volcano Snaefelsjokul. Many of these themes also persisted in the murky field of racial politics. The identification of the Arctic regions with a superior northern European race surfaced in the works of Friedrich Nietzsche, who opened his 1895 essay *The Antichrist* with a metaphorical identification of his new northern supermen with Hyperborean peoples of the Arctic, whom he contrasted with degenerate southerners:

We are Hyperboreans—we know well enough how remote our place is. "Neither by land nor by water will you find the road to the Hyperboreans": even Pindar, in his day, knew that much about us. Beyond the North, beyond the ice, beyond death—our life, our happiness... Rather live amid the ice than among modern virtues and other such south-winds! (Trans. H.L. Mencken, 1920)

Another ancient name from the legendary geography of the Arctic re-emerged in 1912 with the founding of the Thule Society. This mystical German order believed that the island of Thule had been a northern Atlantis, home to an advanced civilization that was destroyed by catastrophe. Some of the secrets of this civilization survived under the guardianship of ancient masters who could be contacted by adepts in the mystical practices of the order. With the help of this knowledge, the Thule Society hoped to create an Aryan tribe of supermen who would eventually exterminate the inferior raccs of the world. The Thule Society served as a recruiting centre for Bavarian radicals, who in turn formed the nucleus for the National Socialist Party organized in the early 1920s by Adolf Hitler. A strain of occult belief characterized the social and political thinking of German Nazism, and elements of belief in a fantastic and legendary Arctic world survive in the racist convictions of white supremacists today. *Thule* is now the title of a venomous American magazine printed in honour of white "prisoners of war" held in American penitentiaries for acts of violence against the non-white races.

A related and equally fantastic belief appears in a number of current publications, and is widely promoted on internet sites devoted to esoteric ideas. Many of these sources are associated with the "Hollow Earth Research Society," whose devotees draw heavily on the ideas of John Symmes. The proponents state with conviction that after the Second World War more than 2,000 eminent scientists disappeared from Germany and Italy, emigrating together with almost a million other people to "a land beyond the North Pole." One version of the tale, which is reminiscent of the Thule Society's beliefs, is based on a widely published "secret log" purportedly kept by the American Admiral Byrd of a flight over the Arctic on February 2, 1947. The "secret log" records the discovery of a tropical landscape and a radiant city, where Byrd was forced to land by flying saucers bearing Nazi swastika markings, and was interviewed by very Germanic-looking space-aliens of great wisdom and power. An alternative version places Admiral Byrd's 1947 flight and discovery of a tropical land in Antarctica. The famous aviator actually was in Antarctica at the time, with a senior command in Operation High Jump, a 1947 U.S. Naval operation designed to provide postwar employment for the navy as well as establish an American presence in Antarctica. The esoteric version states that this venture was in reality an expeditionary force of over 100,000 men sent to Antarctica to destroy the people of the German exodus, who had apparently reached a warm and habitable Antarctica through the hollow interior of the earth. However, because the Americans suffered a humiliating defeat the entire episode was covered up and subjected to a conspiracy of silence enforced by the American government.

This bizarre series of stories combines several elements of legendary Arctic geography. Polar access to the interior of the planet is reminiscent of the *Inventio Fortunatæ*'s description of the polar whirlpool sucking ships into the bowels of the earth, and of Edgar Allan Poe's more recent and (in his mind) independent description of an Antarctic maelstrom. The idea of a tropical refugium in Antarctica, like the rumours of hidden valleys of palm forests in northern Canada and Arctic Europe, has a respectable lineage from at least medieval times, and may in fact be traceable to the mythical ancient Land of the Hyperboreans.

The fact that such concepts still retain a hold on the popular imagination should not be surprising. The legend of the Arctic as a distant paradise has been with us since the time of Homer, and a myth of such antiquity will not fade away merely because of the ephemeral reports of a few Arctic explorers, or the scanning eyes of satellites sent aloft by a single generation.

3 A HUNTER'S WORLD

MY EARLY FASCINATION WITH THE ARCTIC was fed by images of its alien beauty. Like other southerners I saw a land of rock and ice, stripped to its essential stark form by the absence of forests, farmlands and the intrusive remains of human endeavour. The perpetual daylight of Arctic summers added to the disorienting, dreamlike strangeness of this land, while the endless darkness and homicidal cold of winter were alluring in their menace.

I came to know this Arctic through the literature of exploration. For Europeans who found themselves in the grip of the polar regions and who afterwards wrote about it—mostly young men of privileged background who were drawn northwards in a quest for wealth or advancement in their chosen career—the Arctic was a place of infinite hardship and danger. Its beauty lay in its wildness, the absence of the familiar, and the testing of the human spirit in clean combat between intrepid men and relentless nature. Looking back on my early visits to the Arctic, I can see how the writings of explorers provided the framework within which my own perceptions of the region were formed. All southerners suffer "environment shock" in a land that lacks any element of reassuring familiarity. When this shock is received by a mind steeped in the literature of polar exploration, the

HIGHLAND FOREST

With over 20 miles of natural trails, Highland is ideal for hiking, walking, running, horseback riding and mountain biking. In the winter, there is cross country skiing, sledding and snowshoeing. Ski and snowshoe rentals are available.

CROSS COUNTRY SKIING
December-March. Skiers of all levels can enjoy the challenges of a wide variety of groomed trails that feature elevation changes up to 400 feet.

MOUNTAIN BIKING
May-September. Bikers utilize ski trails plus some dedicated biking trails maintained by CNY DIRT.

HORSEDRAWN SLEIGHRIDES
December thru February, 11am-4pm, weekends & holidays (except Christmas) on a walk-in basis. Evening group rides by reservation.

SKYLINE LODGE

This perfect public spot has a fabulous lounge with a fireplace for winter visitors, and Adirondack chairs for relaxing in the fresh air of spring, summer and fall. The concessionaire offers a hearty menu on weekends, holidays and school breaks during the winter. Weddings, family gatherings and business outings can be accommodated year-round.

HOURS & ADMISSION
April-Nov 8:30am-5:30pm
Dec-March 8:30am-4:30pm
Admission payable at Skyline Lodge

1254 Highland Park Rd
Box 31, Fabius, NY 13063
highland@ongov.net
(315) 683-5550

OnondagaCountyParks.com

ONONDAGA COUNTY Parks

County Executive: Joanne M. Mahoney
Commissioner: Bill Lansley
*Fall waterfall photo by Matthew Conheady.
Post Standard photos: skiers by Stephen D. Cannerelli, waterfall by Stanley Walker*
Highland 05/15•15M

PRATT'S FALLS

A 137' waterfall and miles of trails provide solitude, inspiration and beauty.

Camp Brockway
The lodge offers a private setting from late spring to early fall for weddings or family gatherings.

Park Hours
April-Oct • 8:30am-5:30pm

7671 Pratt's Falls Rd
Manlius, NY 13104
(315) 435-5252

To reserve Skyline Lodge or Camp Brockway for your next event, go to OnondagaCountyParks.com and click on Reservations.

FORESTS & FALLS...
SYRACUSE HAS IT ALL!

Highland Forest
& Pratt's Falls

ONLY TWENTY MINUTES
FROM SYRACUSE, NY

SKYLINE LODGE

ONONDAGA
COUNTY
Parks

allure of the Arctic is seen as deceptive: an illusion behind which peril
and adversity are waiting, just beyond the experience of the moment.

This promise of excitement and danger is not part of the experience of the people who are indigenous to the area. Their beloved
homeland is a world of beauty, security and comfort, a world that has
provided a rich livelihood for their ancestors over uncounted generations. These ancestors long ago recognized the essential fact that the
wealth of the Arctic lies in its animals, and that for hunting peoples
the tundra and the ice-covered ocean provide a more easily harvested
supply of animals than do most other regions of the earth. There are
several good reasons why this should be so. The first is that in comparison to the temperate or tropical zones of the earth the Arctic
regions support very few species of animals. This may be due in part
to the harsh climate, but the relatively recent emergence of Arctic
environments from beneath the Ice Age glaciers is also a factor, in
that only a few species have had time to form close adaptations to
the region. The other side of this coin is that a lack of competition
with other animals using the same food supply means that the Arctic
can support vast numbers of individuals of these species. Farmers,
fishers and other commercial harvesters know that such a "monocul-
ture" allows resources to be used very efficiently, and the Arctic
provides this natural advantage to its hunting peoples.

Another feature of the Arctic environment that benefits hunters is
the extreme range of seasonal variation. This circumstance promotes
dense aggregations of animals at certain times of the year, as fish, birds
and mammals take advantage of the brief summer to spawn or to
raise and feed their young, and avoid the cold winter by migrating
southwards or following the edge of the sea ice. Immense herds of
caribou straggle northwards each spring to bear their calves near the
Arctic coast, and then in late summer gather in wide rivers of animals flowing southwards to the shelter of the forests. Seals bear their
pups in innumerable herds on spring sea ice, or in dense colonies on
rocky islands. Migrating whales take advantage of narrow leads in the
sea ice to reach their Arctic feeding grounds as early as possible in the
season, and are funnelled into narrow bands of water where they can
be easily approached. For a few ice-free weeks each summer the

The most sterile reaches of the Arctic support animal life, such as this muskox crossing the gravel beaches of Port Refuge, High Arctic Canada. (Robert McGhee, Canadian Museum of Civilization)

rivers are heavy with runs of char or salmon, and the tundra is dotted with the nests of geese and ducks and swans raising their chicks and moulting in preparation for the early flight south.

To people who have learned when and where to expect these seasonal concentrations of animals in their local environment, how to schedule their yearly activities to make the best use of them and how to take advantage of the winter cold to store meat for weeks or months against seasons of scarcity, the Arctic is a rich and rewarding land. In comparison, the boreal forest directly to the south and the temperate forests to the south of that—where prey animals are more likely to be solitary or to wander unpredictably in small groups, and where the view of the terrain is everywhere obscured by trees—are environments that require a hunter to have a great deal more skill and local knowledge in order to survive.

Early anthropologists who tried to explain why the Eskimos and other Arctic peoples occupied such bleak northern regions tended to suggest that their ancestors had been pushed north by hostilities, or were prevented from moving southwards by aggressively territorial forest populations. However, it seems more likely that the ancestors of most Arctic peoples chose these regions for the prosperous life that they provided. There are drawbacks, of course, even for hunters who learn to exploit the resources of the Arctic. The immense depth of the winter cold and the accompanying darkness and absence of migratory animals create an environment that is less forgiving of mistakes or bad luck than are more temperate regions. Arctic hunters have always lived with less security than those of more southerly countries, but have considered the beauty and wealth of their homelands to more than compensate for its occasional dangers.

Anthropologist Igor Krupnik has made a close study of Arctic hunting peoples, and notes that all have developed similar patterns of life, which are probably necessary for dealing with the inevitable insecurities of Arctic existence. One of these characteristics is having a large number of children, so that local populations have the potential to increase rapidly during times of plenty. No traditional Arctic peoples attempt to limit family size or population growth to the number of humans who can be securely supported by the local environment. Given the instability of Arctic animal resources, an equilibrium between long-lived humans and short-term fluctuations in food supplies is probably impossible. Under traditional conditions, high population growth means that in the inevitable seasons of famine, many people die. But if numbers were smaller, local groups might be wiped out entirely. If there are many, a few will usually survive to perpetuate the group.

Another characteristic of Arctic peoples is a penchant for mobility. When local food sources fail, the capacity to travel far and fast, together with a wide network of kin and acquaintances who can be depended on for temporary support, is of the utmost importance. But Krupnik points out that extensive territorial expansions and long-distance migrations typically occur during times of plenty. Such movements can take place when rapidly increasing local populations expand into areas that had been abandoned during earlier episodes of scarcity. It may also involve moving into previously unknown or unexplored territory in search of good hunting, and rapidly populating such areas when clusters of previously unhunted animals are found.

A third characteristic of Arctic peoples is the predilection for killing as many animals of prey species as possible, with the excess meat either stored for future use, shared with other groups or simply wasted. The casual and routine overkilling of animals on which people depend for their livelihood belies the romantic view of indigenous hunting peoples as natural conservationists, but it is undeniable that it occurs regularly. Overkilling, with no consideration of the biological consequences, has been consistent among Arctic hunters throughout history. In most cases the hunted species have not been seriously affected, since Arctic hunters are few and until the past century they did not possess weapons that were capable of making major inroads into animal populations. However, there is clear archaeological evidence that ancient hunters were capable of causing at least local extinctions. The muskoxen of Banks Island in Arctic Canada appear to have been wiped out at least twice, once about 3,000 years ago and again in the 1850s. The disappearance of the last surviving mammoths, on Wrangel Island far out in the Chukchi Sea, coincided with the appearance of human hunters on the island about 5,000 years ago. Krupnik argues that the extinction of many species of large animals in both Eurasia and North America at around the end of the Ice Age may implicate human hunters, who were living and working in an Arctic-like environment and developing the hunting customs still practised by Arctic hunters of the present day.

Such hunting practices may seem simply reckless or heedless of consequences, but in fact they are intimately tied to the religions and worldviews of Arctic peoples, and these aspects of life are clearly relevant to any understanding of how Arctic hunting societies operate. In their traditional perception of the world and how it works, Arctic hunters generally see no link between overkilling and extinction, or even between overkilling and a drop in the population of prey animals. If any generalization can be made about the wide variety of beliefs held by individual northern societies, it is that the killing of an animal is universally seen as a consensual act between hunter and hunted. The prey presents itself to be killed, selecting a hunter who is sufficiently skilled and knowledgeable as to how the body of the animal is to be treated. If the hunter follows the prescribed ritual

designed to appease or please the soul of the dead animal, the creature is reborn in order to be killed again and provide for the needs
of a hunter who has proven worthy of the kill. Animal populations
decline or disappear not because they have been overhunted, but
because hunters have not treated them with sufficient respect, and
they have decided to avoid these hunters in the future. This view of
the relationship between animals and humans appears to be well
suited to the needs of people who must kill as often as possible in
order to provide against the inevitable times of hunger that every
Arctic hunter has experienced.

The rapport between hunter and hunted is one aspect of an
elaborate belief system that once permeated the lives of northern
peoples and still has considerable power. Anthropologists have used
the term "shamanism" to describe this worldview, from the Siberian
word "shaman" referring to a religious practitioner. Shamanistic
religions were practised by all Arctic peoples, from the Saami of
northern Europe to the Inuit of Arctic Canada and Greenland.
Elements of shamanism can be seen in the old Norse religion
described in the Icelandic Eddas, and scholars have noted striking
similarities to Tibetan Buddhism.

The shamanic view of the universe is much different from that
held by most peoples of the modern world. Instead of a sphere circling a blazing sun, which is itself spinning through endless dark
space, the shamanic world is a stable series of stacked planes. In the
simplest forms of belief there are three planes. The upper is a sky-
world occupied by the sun and moon, planets and stars, clouds, wind
and an assortment of supernatural beings and spiritual forces. Below
this is a midworld occupied by human beings, other animals, ghosts
and wandering souls; this is the portion of the world that is best
understood and most easily controlled. Beneath it lies an underworld
accessible through caves and fissures in the rock and evidenced also
by the fossilized bones and occasional frozen carcasses of huge animals; this mysterious world is inhabited by strange creatures, both
corporal and spiritual, and at times by ghosts. Some societies add an
undersea world, inhabited by fish, sea mammals and powerful spiritual forces. Others insert additional, more esoteric planes, less clearly

envisioned and generally not well understood. Joining these planes and supporting the upper ones is a central world-tree, which serves as a means of communication throughout the universe.

The spiritual forces that occupy the upper and lower planes can have a very significant effect on the occupants of the midworld. Poor hunting, bad weather, illness or individual bad luck are commonly attributed to the malevolence of beings who reside above or below the midworld. In order to appease such beings, a shaman must travel to their home and undertake an apology, a struggle or a threat that will resolve the problem. This is accomplished through spirit-flight, during which a shaman's soul leaves his or her body in order to travel to the sky, the underworld or the world beneath the sea. Spirit-flight can be undertaken only by a person who has been trained in shamanic techniques, or has been selected by the spirits who provide the necessary powers. These powers are often gained through spirit-helpers, most of which resemble an animal and possess a superior form of that animal's abilities: strength, swiftness, cleverness, courage or the ability to see beyond the horizon. Spirit-helpers come to a shaman in dreams or during periods of fasting and deprivation; at times they may be sent by another shaman. The spirit-flight is accomplished in a trance induced either by narcotics or by drumming: the large circular tambourine-drum is the instrument used by all shamans, and is sometimes referred to as "the shaman's steed" or "the door to the spirit-world."

The shamanic religion provides northern peoples with an explanation for the cycles of good and bad times that are inevitable in Arctic life, and gives a sense of control over the disasters of poor hunting, bad weather or illness. It provides small societies with the leadership of shamans who, if they are experienced and wise, may be able to make and enforce difficult but necessary decisions by presenting them as the demands of the spirit-world. Such a position is also, unfortunately, easily abused by unscrupulous individuals in order to gain personal power and wealth in societies where sharing and egalitarian harmony is the ideal. For good and bad, Arctic peoples have held to the shamanic view of the world for thousands of years, and some of its tenets have proved highly resistant to the forces of

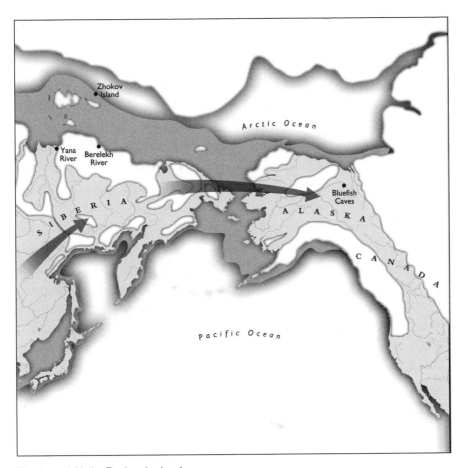

Siberia and Alaska During the Ice Age

both missionary Christianity and the science-based administration of hunting regulations.

The complexity of Arctic worldviews and adaptations implies a great time-depth for human occupation of the polar regions. However, this does not mean that any Arctic society has simply lived and developed in its local area since time immemorial, or since the end of the Ice Age. The romantic notion of ancient and unchanging cultures is very much at odds with what we know of Arctic societies. The small nations of the north have histories that are as long and as complex as those of more southerly peoples, a fact that archaeological findings are continually and surprisingly demonstrating.

The problem with archaeology is that the discoveries it makes are of randomly preserved occurrences, and it sheds only narrow and scattered beams of light into a large and murky expanse of time and space. A couple of such finds, mentioned in Chapter 1, seem to indicate that humans had penetrated beyond the Arctic circle to the valley of the Yana River and across Beringia to reach northwestern North America as early as 25,000 to 30,000 years ago, at the height of the last ice age. However, there is nothing to link these finds to the next set of discoveries, which are only about half that age, dating to the period when global temperatures began to rise and the Ice Age came to an end. These sites are on the Berelekh River, a northern tributary of the Indigirka, where hunters left ivory tools and ornaments in association with a large collection of mammoth bones dating to about 12,000 years ago; and in a cave above the Bluefish River in the northern Yukon, where at about the same time another group of hunters left a few stone tools. We know nothing else of these people, and again there is a gap of several millennia until the next well-documented finds, which probably relate to new sets of peoples moving northwards into the tundra and along frozen coasts.

By about 8,000 years ago the continent-spanning glaciers of the last ice age had retreated to the remnants that still exist today, and the environmental zones of the present world had become established in an approximation of their modern forms. There is widespread evidence that by then people were moving into at least the fringes of the Arctic regions, primarily adapting coastal ways of life developed in more southerly regions. Some of the rock art of northern Norway may be 8,000 years old, and its painted images of Arctic animals and hunters tell us more than the few collections of stone tools that date from this period. Across the Atlantic, people on the tundra coast of Labrador were hunting walrus and burying their dead beneath mounds of boulders with rituals as complex as any known to have been used elsewhere in the world at that time. Maritime hunters had advanced northwards along the Pacific coast of Asia, at least as far as Kamchatka and perhaps to Chukotka, while on the American side, South Alaska and the Aleutians supported coastal hunting societies.

We know significantly less of early ventures into interior areas, where archaeological sites are more widely scattered and harder to

The limestone outcrop containing Bluefish Cave overlooks a broad valley in the northern Yukon, just to the south of the treeline. (Jacques Cinq-Mars, Canadian Museum of Civilization)

find. We know that by 8,000 years ago caribou-hunting peoples occupied much of interior Alaska, and were making at least seasonal forays across the Barren Grounds west of Hudson Bay to reach the Arctic coast. Similar ventures were being made across Siberia, but until recently we knew very little about the extent of Arctic adaptation there. This picture changed remarkably over the past decade with the discovery and excavation, by archaeologist Vladimir Pitul'ko, of a large and deeply frozen archaeological site on tiny Zhokov Island. This island is part of the Novosibirskiye Archipelago, 500 kilometres north of the Arctic coast and in a region that is so isolated that it was thought to have been unoccupied until relatively recent times. However, radiocarbon dates indicate that by about 8,000 years ago, at a time when the island was on the coast of the Siberian mainland, people were living in wooden houses three or four metres across, excavated into the ground and roofed with driftwood. They hunted reindeer and polar bears as well as a few seals and walrus, used sleds similar to those of much later times, and kept dogs, which were probably used to pull the sleds. The people who lived in this settlement had developed a true Arctic adaptation and, at 76° North, the Zhokov Islanders had obviously learned to cope with winter cold and darkness as severe as that of most polar regions.

Whoever the Zhokov Islanders were, we cannot trace their descendants to any peoples of the present world. Neither do we know whether the artists who left the early rock paintings of northern Norway were ancestral Saami, or whether the early hunters of Alaska were ancestral Eskimos, Aleuts or Dene Indians. Given the tendency of traditional northerners to rapidly expand the size of their local group during times of plenty, to suddenly move into new and distant hunting grounds, and to face extinction or near-extinction in inevitable famines, it seems likely that many of the peoples who left archaeological evidence of early Arctic occupations have no descendants in the present world. The best example of such a people are the Tuniit, the first widespread occupants of Arctic North America, and a people who left a treasury of well-preserved art depicting a very distinctive shamanic view of the world.

By about 8,000 years ago the Arctic environments of North America were as extensive as they are today, and animal populations had moved northwards to establish themselves on lands and in sea-channels recently freed from glacial ice. Although ancestral Indian groups made summer excursions northwards across the tundra, probably following the caribou as Dene and Innu groups have done in recent centuries, it seems likely that they retreated in winter to the shelter of the forests. There is no indication that any society had developed a year-round adaptation to life on the tundra, and for the following 3,000 years most of Arctic North America remained unoccupied.

Then, at some time around 5,000 years ago, the picture suddenly and mysteriously changed. Archaeological remains dating from between 5,000 and 4,500 years ago are found everywhere along the western and northern coasts of Alaska, across Canada's Arctic archipelago and as far as the barren valleys of northern Greenland. Moreover, these remains are very uniform, very distinctive and look like nothing that has been found belonging to earlier periods in the northern forests or subarctic coastal regions of North America. What they do resemble are the archaeological remains left by Siberian peoples from Lake Baikal to

Chukotka: tiny stone tools chipped skilfully from brilliantly coloured flints; minute needles honed from bird bone, their drilled eyes almost microscopic; harpoon heads carved from antler or ivory, with a hole to which the hunter attached a line to retrieve his prey from the sea; bone or ivory lances edged with razor-sharp flint blades.

Their dwellings appear to have been as distinctive as their tools and weapons. Most of the vestiges found look like the remains of tents; oval patterns of boulders were used to weight the edges of the skin cover, which was probably stretched over a conical framework of poles and deeply banked with turf and snow for winter insulation. The ground-plans are small, usually two to three metres on a side, and are divided by a central passage that runs from front to back. This unique feature is usually about eighty centimetres wide, edged with rocks or upright slabs, sometimes paved with flagstones, and contains a central hearth. Like the styles of tools and weapons, the midpassage dwelling resembles those used by Siberian peoples, not only 5,000 years ago but up to the past century, and by societies as distant as the Saami of northern Scandinavia.

The sudden appearance of this distinctive way of life across the North American Arctic must reflect a movement of newcomers from Siberia, peoples who either crossed Bering Strait by boat or, more likely, walked across on the dangerously shifting winter ice. Their Arctic way of life was probably made possible by the Eurasian inventions that they seem to have introduced to the Americas: the bow and arrow, and finely tailored skin clothing of the type worn by Siberian peoples up to the present day. The arrows are evidenced by tiny chipped stone points with sharp penetrating tips; the clothing by the vast numbers of bone needles, scraps of sewn skin, and by occasional carvings showing individuals dressed in a costume consisting of high boots, trousers and a double-layered parka with a high collar. With sophisticated clothing, together with the harpoons, lances and fish-spears that allowed them to exploit most Arctic resources, and the small tents that provided portable shelters from the worst of Arctic weather, these people had everything they needed to travel and hunt on tundra and sea ice. Curiously, archaeology has not found much

The remains of a 4,000-year-old stone hearth on a beach in the Canadian High Arctic marks a camp of the early Tuniit, the first humans to learn how to live in this extreme northern environment. (Robert McGhee, Canadian Museum of Civilization)

evidence of other typical elements of Arctic adaptation: dog bones have been found very rarely, indicating that like all other hunting peoples they were accompanied by canine colleagues, but also suggesting that these were very few. The wooden parts of a single kayak-like boat have been found at a site in Greenland, indicating that these people knew how to make and use watercraft, but again suggesting that few such useful appliances existed.

This small ivory maskette, portraying a serene and tattooed face, is the earliest known portrait of an Arctic individual. It was carved about 3700 years ago by a Tuniit artist on Devon Island, Arctic Canada. (Richard Garner, Canadian Museum of Civilization)

Who were these first Arctic explorers? Archaeologists call them "Palaeo-Eskimos" (Old Eskimos), simply because the country that they occupied was much the same as that occupied by Eskimo-speaking peoples of the past few centuries. The name is misleading, since it is very unlikely that these people spoke an Eskimo language, or were direct ancestors of today's Eskimos or Inuit. A far better name is that which was given to them by the Inuit themselves, who recognized them as a distinctive and strange race. Inuit oral history

relates that when their ancestors arrived in Arctic Canada a few centuries ago they found the country occupied by a tribe of large, strong
and inoffensive strangers whom they called "Tuniit," and whom the
ancestral Inuit soon drove from their lands. The Tuniit are almost
certainly the people whose archaeological remains show almost
5,000 years of occupation in the Arctic regions of North America,
from Bering Strait to the east coast of Greenland and south to the
island of Newfoundland.

I was introduced to the Tuniit in the summer of 1965, when I
finally reached the Arctic. I was working as a field assistant to Bill
Taylor, an archaeologist at the National Museum of Canada who was
soon to become a distinguished director of that institution. For three
months we camped and worked on the banks of the Ekalluk River
in Canada's central Arctic, uncovering the remains left by almost
3,000 years of occupation. The Ekalluk is a brilliantly clear, cold tor-
rent only a few kilometres in length, draining a large lake that
extends into the interior of Victoria Island. It is named after the
Arctic char, *iqaluk* in the language of the Inuit, and vast numbers of
these fish transit the river each summer in their short migration from
lake to ocean. For millennia it has also served as a crossing-place for
migrating caribou funnelled to this narrow isthmus between lake and
seashore, and for much of that time the spot has attracted fishers and
hunters eager to make use of its resources.

We began our excavation in the tundra meadows that extended,
lush with flowers, back from the river. Almost everywhere we dug
we turned up the remains of an ancient presence: small stone
weapon-points and knife-blades and skin-scrapers chipped from
pink quartzite; elegantly designed harpoon heads and fish spears
carved from antler and stained a glossy chocolate by burial in the
frozen tundra soil; everywhere the roots of grass and flowers penetrated thick layers of broken caribou bones. Across the river lay
several large outlines of stone, the remains of ceremonial enclosures
where Tuniit families had joined together to celebrate the success of
yet another hunt.

During that summer I became fascinated with the challenge of
attempting to understand the Tuniit. I was intrigued by the simple

elegance of their tools and weapons, tiny artifacts so skilfully made and adhering so carefully to standardized patterns that they almost seemed the products of industrial design rather than home crafts- manship. Against this stood the apparent inadequacy of this technology in comparison with that of the Inuit who later inhabited the area. Where were the many gadgets used by the Inuit to capture a variety of animals, to process meat and skins, to test and cut snow for construction, to trim oil-lamps and carry out dozens of other tasks? Where were the remains of boats, dogsleds, even of adequately insulated winter dwellings?

My bewilderment only increased in later years when I had oppor- tunities to get to know more of the Tuniit. In the polar desert of the High Arctic islands I encountered the remains of a very early and dis- tinctive Tuniit society, which the Danish archaeological pioneer Eigil Knuth had named the Independence culture after the major fiord system of that name in northern Greenland. The remains of Independence dwellings were simple stone outlines of tents with central midpassages enclosing a box hearth, and the small numbers of bones and artifacts found in their vicinity indicate that they were typically used for only a few days or perhaps weeks. These early Tuniit—Independence sites often date as early as 4,500 years ago— seem to have been bow-hunters of muskoxen. They apparently did not use or hunt the marine mammals on which later peoples depended for the oil used both in their diet and in the lamps sup- plying winter light and heat. No stone oil-lamps have been found in their sites, nor the remains of any closed and insulated dwelling that could have been heated with such lamps. Eigil Knuth proposed that these first venturers into the extreme High Arctic, where the winter night lasts for months, passed the entire year in tents covered with muskox hides and heated with small open fires. The hearths of their tents usually contain charred fragments of muskox bone and pencil- sized Arctic willow twigs, but in the polar desert such fuel can have provided only occasional fires. Knuth was tempted to picture fami- lies surviving the long winter night in dark, unheated tents, perhaps in a state of torpor mimicking the hibernation by which some mammals survive northern winters. A way of life such as this would

This 18th century illustration of a Saami household from northern Europe shows an arrangement of midpassage and central hearth, demonstrating an ancient continuity among the peoples of the Arctic world. (From Knud Leem, *Beskrivelse over Finmarkens Lapper.* Copenhagen: G.G. Galifarb, 1767)

have been considered intolerable by the Inuit or by other traditional Arctic peoples, and we may be mistaken in reconstructing such a bleak, meagre and uncomfortable mode of living. On the other hand, it seems possible that at least some tribes of Tuniit accepted, and learned to contend with, living conditions beyond the range of those known to any peoples of the recent world.

A few years later on tiny Dundas Island, a fogbound scrap of grey rock surrounded by the ice floes and swirling currents of a High Arctic polynia, I encountered another Tuniit society. These people had lived only about 1,000 years ago, and had developed the capability of hunting marine mammals as large as the walrus whose grunts and roars echoed continuously through the fog. They had built their small village—half a dozen tent foundations with mid-passages and central hearths—in the lee of a sea-rounded outcrop of rock, and this shelter had filled with drifting snow the following winter. Each year through the centuries the huge snowdrift had returned, and every summer it melted back to reveal the site for only a few days in late August, so that when I visited the village it was as completely preserved as if it had been abandoned the previous month. Moss mattresses covered the sleeping areas on either side of the midpassages, chips and curls of wood lay where they had been whittled, and in one dwelling the housewife had covered the hearth with a flat rock to preserve the charcoal as fuel for her next visit. I had the sensation on this August day that I was not alone; that at any moment there would be the sound of voices, of feet moving through gravel, and the occupants of the village would appear from the fog. I was somewhat uneasy collecting their possessions for my own use, and especially those belongings that had obviously been more than simple tools: a polished ivory carving of a plump ptarmigan; a small wooden figure representing a man with a hole in his chest, the hole containing a splinter of wood; and, deep in a crevice in the rock behind the village, a small wooden carving of a bear.

These later Tuniit were a part of the development that archaeologists call the Dorset culture, named after the Baffin Island community of Cape Dorset where their remains were first recognized. Between about 500 BC and AD 1500 the Dorset people were expert hunters of seal and walrus, burned the oil of these animals in stone lamps as winter light and fuel, and must have had a considerably more secure and comfortable existence than that of their earlier ancestors. Most strikingly, they took the elegant design, miniaturization and careful detail of their ancestors' technology and applied it to the creation of minute sculptures. Humans, all of the animal species of the Arctic world, beings in transformation between animal and human states, and other human-like creatures were portrayed in ivory, antler, wood and occasionally soapstone. The standardization of many portrayals shows that these are not works of the individual imagination, but are rooted in a singular set of religious beliefs and views of the world.

Some of the objects found in Dorset villages are directly related to the activities of shamans: wooden frames of small tambourine drums; life-sized masks carved from driftwood and painted with red ochre to portray faces that are not quite human; carved sets of ivory teeth with protruding canines that could have been slipped into the mouth to transform a human into a bear. Small ivory tubes have also been found, similar to those used by shamans in other northern cultures to suck magical weapons from the body of a sick patient; the Dorset tubes have one end modified to form a gaping mouth flanked by two animal heads—always a wolf and a caribou, hinting that some form of power might emanate from the antagonism of these perpetual enemies. Tiny ivory harpoon heads, a few millimetres in length, suggest the magical weapons sent by shamanic witches to cause illness. Small sculptures of humans or animals with holes in the throat or chest may have been used as sources of magical power over the creatures depicted. Carvings of bears portray the animals with trailing arms and legs, as if they were floating or flying; the skeleton is visible on the surface of the bear's body, reminding us that in shamanic thought the skeleton and the soul are closely associated as the two elements of the body that survive death. Such a flying bear or bear-spirit may represent the powerful helping spirit of a shaman, or perhaps the shaman himself while on a spirit-flight to another realm. Wands of caribou antler are covered with carved images of human and human-like faces. One such baton that emerged from the frozen soil of a village I was excavating portrays sixty faces, some of which have distinctly animal-like features, and a portion of the wand is heavily worn as if it had been habitually carried in the hand. This object may depict the spirit helpers of a shaman, or perhaps simply a series of caricatures of the various members of the local group.

The art of the Dorset people is so varied and intricate that it allows a glimpse of the spirit-world known to these ancient hunters, a world resembling in many ways that of other northern shamanic peoples, yet unique to this society that developed in the relative isolation of Arctic North America over a period of almost five millennia. Their art also appears to foreshadow the end of Tuniit society. In the last few centuries of their existence, between about AD 1000 and 1500, Dorset artists produced an ever-increasing number of amulets and objects of shamanic use. This was also a period of increasing stress on the societies to which the artists belonged. In these centuries the northern hemisphere was subject to the warming climatic conditions that are known in Europe as the Medieval Warm Period. To Tuniit hunters who seem to have been adapted primarily to hunting on land and sea-ice, the unexpected appearance of open water during early summer or a delay in the expected freezing of autumn seas would have brought hardship and often disaster. Inuit today worry that the warming seen over the past decade will bring disastrous changes to their way of life, and the situation a thousand years ago would have been as serious.

At the same time, the Tuniit world began to be visited by outsiders, people whose existence could not have been even suspected by the shamans on whom the Tuniit depended for their wisdom and knowledge. From the east came the Greenlandic Norse, coasting the shores of the Tuniit on their explorations towards the country they called Vinland. Later, as archaeologist Pat Sutherland has recently demonstrated, some of these strangers made lengthy visits ashore and probably engaged the Tuniit in trade. The visitors must have brought metal and other exotic items that they would have been willing to exchange for walrus ivory, and they may also have brought, more insidiously, viruses or bacteria to which the isolated Tuniit had no biological resistance. The faces of the newcomers—long narrow faces with pronounced eyebrows and noses, sometimes with a beard or cap—are portrayed on antler wands carved by Tuniit artists of the period.

An equally unexpected and ultimately more serious challenge to the Tuniit homelands soon began to arrive from the west. Ancestral Inuit penetrated the Eastern Arctic during the twelfth and thirteenth centuries AD, perhaps attracted to the region by the availability of European trade-goods brought by the Norse. These Inuit were

whale-hunters from the large coastal villages of northern and western Alaska, entrepreneurs who for generations had controlled the
growing iron trade across Bering Strait. With their dog-sleds and
umiaks—large skin-covered boats that could transport the people,
dogs and equipment of an entire camp—they were capable of
exploring and occupying vast stretches of land. They brought with
them the Mongol type of recurved bow, a weapon that was as effective for hunting humans as it was for animal prey. They also brought
a tradition of savage intercommunity warfare that had been honed
for centuries in coastal Alaska. Unlike Norse farmers, who had little
interest in settling Tuniit lands, the Inuit would have seen the region
as a bountiful country that they could occupy and exploit in comfort and security. To such a people, the small, scattered bands of Tuniit
can have been seen only as pitiful and annoying competitors who
could easily be driven from the lands that the Inuit needed for themselves. Inuit oral tradition tells us that this was what happened again
and again as the Tuniit—large and strong people who nonetheless
lacked the weapons and the fighting spirit of the Inuit—were either
killed or moved away to an unknown fate. Only one story provides
a glimpse of how the Tuniit felt about the situation: in 1922 the Inuit
historian Ivaluardjuk told anthropologist Knut Rasmussen of the last
Tuniit who lived on the Uglit Islands in Foxe Basin:

> The Tuniit were a strong people, and yet they were driven from their
> villages by others who were more numerous, by many people of great
> ancestors; but so greatly did they love their country, that when they were
> leaving Uglit, there was a man who, out of desperate love for his village,
> struck the rocks with his harpoon and made the stones fly about like
> bits of ice.

Inuit tradition also tells that one band of Tuniit survived the anni-
hilation. These were the people who came to be known as
Sagdlermiut, a small and isolated society that occupied Southampton
Island in northern Hudson Bay. Until the early twentieth century
they used tools chipped from stone in the Tuniit manner, instead of
the metal tools employed by other Inuit. Lacking kayaks they ven-

tured to sea on inflated sealskins. They spoke a strange dialect, wore clothes cut to strange patterns and had little to do with neighbouring Inuit groups who considered them too backward to be of any interest. Recently, biological anthropologist Geoffrey Hayes used DNA recovered from Sagdlermiut skeletons to confirm that the genetic pattern of the Sagdlermiut resembled that of Dorset people as much as it did Inuit, suggesting that both groups had contributed to their ancestry. Unfortunately, the Sagdlermiut did not survive the next major round of contact with outsiders who intruded into their homeland. In 1902 Southampton Island was visited by the Scottish whaler *Active,* several of whose crew were ill with an unidentified disease. The illness spread to the Sagdlermiut, and by the autumn of that year the entire band had died. After 5,000 years, the Tuniit had become extinct.

This small, wretched group, coughing themselves to death in unheated tents on the gravel beaches of Southampton Island, were the last of the ancient hunting peoples whose way of life had developed in the millennia after the last ice age, in relative isolation from more populous and more technologically advanced peoples from the south. With them died a unique vision of the Arctic world, a vision that we can only glimpse from the tiny and powerful sculptures preserved in the permanently frozen earth of their homelands.

4 IN ARCTIC SIBERIA

THE OLD ANTONOV 24, nearing the end of its twice-weekly flight from Anadyr, broke out of low cloud over the Bering Sea. Wheels and wing-flaps clunked into place, the whine of turboprops dropped in pitch, the loosely bolted seats and the piles of freight lashed at the front of the cabin began to vibrate as we sank towards grey water lined with whitecaps. Suddenly we crossed a gravel beach, then an expanse of grassy tundra littered with oil drums and rusted chunks of abandoned machinery. The airstrip lay in the centre of the small town, and a confusing panorama of decrepit buildings and smoking metal stacks scrolled past the window as we touched down. My seatmate, a Chukchi teenager dressed in jeans and ski jacket, stirred from sleep for the first time since we came aboard, and stared out the window at home. We had landed in Lavrentia, the most northerly community on Asia's eastern coast, and less than 100 kilometres south of Bering Strait.

This was my first visit to Russia after the collapse of the Soviet state, an event that opened many previously forbidden regions to Western visitors. For the first time in the twentieth century, outsiders could freely travel this vast segment of the Arctic world, and experience for themselves a land that had long been inaccessible. When I climbed down from the plane in Chukotka I was confronting not

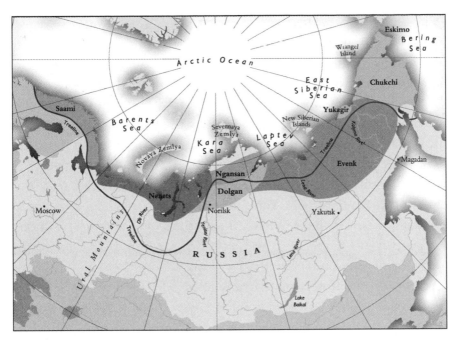

Aboriginal Homelands of Northern Eurasia

just the bitter wind blowing off Bering Sea, but the dark preconceptions that had accumulated in my mind while watching from the safety of Canada as Stalin exiled entire regiments of military prisoners, entire classes of citizens and occasional small nations to the ice and mud and darkness of this unknown land. I had read *A Day in the Life of Ivan Denisovich* and *Gulag Archipelago,* and the mere names Magadan or Kolyma brought a pang of dread at the thought of so much human misery. Siberia was the lair of the nuclear missiles targeted on Western cities. Its impenetrable fastness concealed military laboratories where isolated communities of scientists plotted the destruction of our societies. It was a land where ruthless and reckless industrial development had produced a nightmare landscape of polluted rivers, contaminated soils and radioactive ambushes.

Westerners refer to the entire vast region of tundra and taiga forest, from the Urals to the Pacific, as Siberia; the Russian use of the name is more restrictive. And like most Westerners, and some Russians, I had conventional ideas about Siberia that can be traced to centuries of vague tales told by generations of Tsarist and Soviet exiles—and even further, to an ancient heritage going at least as far back as Marco Polo's account of his thirteenth-century journey to China. During his long residence at the court of Kublai Khan, Polo had heard two stories about the lands to the north. He reported that "... far away to the north of the kingdom, there is a province called Darkness, because there it is constantly dark; the sun, the moon and the stars never appear there, but it is always as dark as with us in the twilight." An adjacent region was described as a land of mud and ice where the inhabitants lived in underground houses on account of the cold, and for lack of roads travelled in sledges pulled by dogs "little less than a donkey in size." Even in the thirteenth century, however, the peoples of Arctic Siberia were not isolated from the outer world. The inhabitants of both Darkness and the Land of Mud and Ice were reported to be hunters whose furs attracted traders and marauders from the south. Yet although their furs were coveted by the Mongol court they remained free peoples; according to Marco Polo, they owed no allegiance to the Mongols or indeed "to any foreign Lord."

The inhabitants of Chukotka and adjacent regions continued to rule their own lands for almost five centuries after Polo's visit to China, a period during which the empire of the Mongols disintegrated and the new Russian empire expanded eastwards across the broadest continent on earth, driven by a coincidence of state policy and commercial interest. By the fourteenth century, traders from Muscovy had crossed the Urals and penetrated as far as the banks of the Ob, the most westerly of the great rivers that rise in the mountains of central Asia and flow northwards to the Arctic coast. To the east of the Ob lay 1,000 kilometres of forested plain stretching as far as the Yenisei, then a further 4,000 kilometres of upland plateaus and mountain ranges, broken by the valleys of the Lena, Indigirka, Kolyma and hundreds of smaller rivers.

Long before the Russians arrived, this vast region was fully occupied by a patchwork of small nations. The more southerly followed ways of life similar to those of the herding peoples of central Asia, semi-nomadic societies based on caring for droves of horses, sheep and goats. Most of these peoples had long paid tribute to the empire of the Mongols and to its predecessors, and in turn exacted tribute from the poorer and weaker forest tribes to the north. The culture of the steppe herders had long since penetrated the forested taiga, where cold temperatures and lack of grazing prevented the maintenance of the domestic animals of the south. Here the economy was based on reindeer, the same species (*Rangifer tarandus*) as the caribou that has long provided the basis of life for most interior hunting peoples of northern North America.

The reindeer supplies most of the basic needs of northern forest people: meat and fat for nourishment; antler to replace the hard-woods available to most other technologies; lightweight and superbly insulated skins that are unsurpassed material for Arctic clothing; and long sinews with which the clothing can be sewn. From Karelia to the Pacific, most of the northern indigenous peoples encountered by the Russians lived with and by the reindeer. Some hunted wild reindeer, some combined hunting and herding, while others lived primarily on domesticated herds. Some herding groups milked their domestic reindeer to provide a nourishing and dependable food supply, while others used them for pulling sleds, as pack animals and mounts for riding and as decoys to attract and trap wild reindeer. The domestication of reindeer has a long history (it is mentioned in the ninth-century account of the Norwegian Ottar, and is alluded to in Marco Polo's description of thirteenth-century Siberia), but it has been argued that the practice became prevalent in Arctic regions only during the eighteenth century, either because wild herds had been reduced by overhunting and disease, or because of a growing market for reindeer hides.

Fishing provided most Siberian peoples with an additional source of protein, and they also trapped the fur-bearing animals of the forest. Sable, marten, beaver and fox furs were the currency of the forest peoples in dealing with the foreign traders who brought iron needles, metal cooking pots and other useful goods. Furs were also the currency of tribute paid as protection from attack by the more populous nations that lived upriver. This system extended through several steps from the peoples of the tundra southwards as far as the Mongol Empire, and to the smaller but equally dangerous kingdoms into which it gradually disintegrated.

Nenets reindeer herders move their animals through winter snow on the Siberian tundra.
(Bryan and Cherry Alexander)

As in seventeenth-century Canada, furs were the magnet that
attracted Europeans to the forests of northern Asia, and as in Canada
royal grants and charters to exploit this profitable resource were
eagerly sought. By the latter part of the sixteenth century the wealthy
Stroganoff family had been granted a monopoly on trading in the
lands adjacent to the Ural Mountains. This region was still subject to
tribute from Kuchum, the Tartar "Khan of Siberia," who also led
raids into eastern Muscovy as far as the town of Perm. The beginning of Russia's conquest of Siberia can be conveniently dated to
1581 when Maxim Stroganoff hired the Cossack Yermak Timofeyev
to protect his trading empire. In language reminiscent of that which
justified medieval crusades, Yermak was ordered to overthrow the
infidel Khan and capture his capital at the town of Sibir on the
banks of the Irtysh River, a western tributary of the Ob. With a few
hundred Cossacks and mercenaries, the latter recruited from prisoners taken in Muscovy's battles in eastern Europe, Yermak crossed the
Urals and accomplished his task in 1584.

Strogonoff had undertaken this endeavour without notifying the

Tsar, and Ivan the Terrible threatened grave retributions if the adventure precipitated further Tartar attacks on eastern Muscovy. Yet all was forgiven when messengers laden with furs arrived in Moscow with news of the victory. Ivan dispatched Prince Bolkovsky with 500 infantrymen to consolidate the conquest, ten priests to establish Christianity in the new territory and valuable gifts to purchase the loyalty of local rulers. Yermak was given the title Prince of Siberia and was commanded to govern the conquered region, which he did by imitating the Tartar practice of exacting tribute from local tribes and destroying those that did not submit.

Yermak Timofeyev's conquest of the Khanate of Siberia has become something of a heroic national myth for generations of Russians, somewhat like the stories of Daniel Boone and similar figures of American frontier legend. According to the tale Yermak took vows of chastity, unfailingly trusted the Christian God in the face of overwhelming odds, was valorous in combat and treated defeated enemies with chivalrous justice. This reputation is at odds with what we know from other sources about the Cossacks: hard-living militaristic tribes of Russians from the valleys of the Don and Volga, whose primary goal was winning fame and riches while retaining their liberty from serfdom.

Cossacks were the explorers, the venturers, the military forces and the traders who pushed Russian influence inexorably eastward across Siberia, lured by the promise of wealth that lay in the furs, mammoth ivory and other resources of the east. By 1628 they had established a fort at Krasnoyarsk in the headwaters of the Yenisei, and four years later at Yakutsk in the middle valley of the Lena. Yakutsk soon became the centre of Russian influence in eastern Siberia, and of further exploration to the Pacific coast. From Yakutsk a band of Cossacks reached the Kolyma, and from there Simon Dezhnev continued eastwards to pass through Bering Strait in 1648 and reach the mouth of the Anadyr where it flows into the Bering Sea. By the end of the seventeenth century Russian influence was entrenched in most of Siberia, with trade and tribute flowing westwards from as far away as Kamchatka and the Amur River. In the following century they expanded to the Aleutians, Alaska and as far as the coast of California.

In Siberia, only the Chukchis and Eskimos of Chukotka remained free from Russian rule until, after a series of bitter and costly battles, they too finally became subjects in 1789.

Peter the Great's 1697 census of Siberia found a population of roughly 150,000 living east of the Urals, of whom about half were ethnic Russians and the remainder aboriginals. Most of the Russians were concentrated in western Siberia, in the valleys of the Ob River and its tributaries, but Russian influence radiated from a growing number of settlements scattered along the rivers and other transportation and trade routes. Siberian furs soon became a significant export from Russia to Europe and Asia, and an important addition to the Russian treasury. In order to maintain the flow of furs the Russian Tsars levied a tax, or tribute, called the *yasak*, on each of the small Siberian nations that were incorporated into the Empire. Although the *yasak* itself was not onerous for many Siberian peoples, the manner in which it was collected—through a tax-farming system allied to grants of trading monopolies—left it open to corruption and flagrant abuse by local merchants and officials. Cases of extortion, of hostages taken to ensure payments and of local revolts put down with brutality by the alliance of mercantile and military powers, were incessant.

The need to obtain furs in order to pay the *yasak,* as well as to trade for iron tools, cooking pots, guns and other useful items, put a new burden on the reindeer-herding peoples of the Siberian forests. Further pressure was added by the spread of Russian farmers across southern Siberia, displacing aboriginal peoples northwards into the forests and disrupting the traditional boundaries between nations. During the eighteenth century the unorganized eastward trickle of Russian peasants escaping serfdom was significantly increased by a flow of political and religious exiles—notably the Old Believers, who established their austere religious communities in several regions. The discovery of gold and other resources usually led to the complete displacement of local occupants in favour of Russian and foreign venturers. As the Russian population increased, Russian hunters and trappers began to compete directly with the natives, paying little heed to traditional rights to the use of the land and its products.

By the early twentieth century, all of the reindeer-peoples of Siberia were involved to some extent in the Russian economy, but for most this was merely an extension of their relationship with earlier peoples of the southern steppes. The building of the Trans-Siberian railway consolidated Russian economic and administrative influence across the continent, but the actual authority of the Russian state decreased rapidly as one travelled away from the main transportation routes and economic centres. Most peoples of the tundra and taiga continued to follow traditional ways of life, which had always accommodated commerce and the paying of tribute for protection from hostile outsiders. Nevertheless, increasing accessibility, combined with a weak and ineffectual Russian administration, led to widespread abuses of local peoples at the hands of the predatory individuals who thrive in frontier situations.

Throughout the centuries of Russian expansion across the great snow-forests, the peoples of the Arctic tundra remained relatively untouched. Their geographic isolation protected them, but they were also bypassed because their countries were not rich in sables and the other fur-bearing forest animals that were the magnet for Russian interest. Approximately 30,000 people hunted and herded reindeer across the tundra plains fringing northern Siberia, or hunted sea mammals from the seasonally frozen coasts.

Those of the western Arctic—the Saami who occupied the tundras adjacent to the Barents Sea and the Nenets of the Kara Sea—first made contact with Europeans who arrived by sea, not with Cossacks entering their country from the interior. This process began as early as the ninth century, when the Norwegian adventurer Ottar told the English King Alfred that he was the first European to have rounded North Cape and explored the entrance to what is now the White Sea. Many sixteenth-century English and Dutch attempts to discover a sea-passage to the north of Asia came to grief on the shores occupied by Saami and Nenets hunters, but their reports of a region rich in sea mammals attracted European walrus hunters and whalers. Russian Pomors, a maritime people, expanded their hunting and trapping activities around the coasts of the Barents and Kara seas during the eighteenth century, probably displacing several indigenous groups that had previously hunted there. The decimation of whale and walrus stocks, which is well documented for Svalbard, also occurred along these coasts so that by the nineteenth century, when the Pomors retreated to their White Sea base, the northern tundras were in the hands of reindeer herders and the coasts were practically abandoned.

Far to the east, the Chukchi and Eskimo occupants of Chukotka remained beyond effective Russian influence until well into the twentieth century. Although the Chukchi had nominally become subjects of the Tsar during the late eighteenth century, they and their Eskimo neighbours along the coast of Bering Sea were the only Siberian peoples exempt from the *yasak* tribute. The Russian government considered the tax that could be obtained from these relatively poor and fiercely independent tundra peoples not worth the cost of collecting, and missionaries found little interest in Christianity among the followers of the shamanic religion that had protected them for uncounted generations. In addition, the new and profitable sea routes from Kamchatka and the Aleutians to Russian America bypassed the Chukchi Peninsula, so its inhabitants were left to develop their own trade with Alaskan Eskimos across Bering Strait. The connections with the Americas were solidified when American whalers and traders flooded the Bering Sea during the latter half of the nineteenth century, followed by prospectors and miners overflowing from the goldfields of the Yukon. I have been told by Chukchis that their great-grandparents spoke English, which was the language of trade throughout the region until the time of the Soviet revolution. By that time the local stocks of bowhead whales and walrus had been so diminished that commercial hunting was no longer profitable, and most American miners and traders had also left the region.

The first years after the Soviet Revolution engulfed much of southern Siberia in the bitterness of civil war, a struggle that intensified the exploitation and neglect of aboriginal populations. With the war won, the new Soviet government began to expand its influence into the farthest reaches of the continent, and for the first time in their history the indigenous peoples of northern Siberia began to

The midnight sun paints the immense snow-covered landscape of northern Ellesmere Island, Arctic Canada. The low angle of the Arctic sun combines with cold clear air to produce lighting effects that are unique to the polar regions. (Patricia Sutherland)

Sun-dogs and parhelia, caused by light reflecting from ice crystals in the upper atmosphere, are among the ethereal effects that characterize Arctic skies. (Bryan and Cherry Alexander)

The ghostly lights of the aurora borealis swirl over Vatnajokull, the largest ice cap in Iceland. (Ragnar Th. Sigurdsson)

This map of the polar regions was published in 1606 by Gerhard Mercator, the leading European cartographer of his day. The concept of the polar region is based on fantastic ideas surviving from mediaeval times: a massive polar mountain emerges from the centre of an open sea that drains to the earth's interior. The magnetic pole is shown as a mountain of iron in the sea between Asia and America. (Royal Ontario Museum)

Caribou during their northward migration across the Barren Grounds to the west of Hudson Bay. Such vast concentrations of animal life make the Arctic a hunter's delight. (Fred Bruemmer)

A tiny arrowhead skillfully chipped from pink flint by an early Tuniit hunter has lain on the surface of this High Arctic beach for the past 4,000 years. (Robert McGhee, Canadian Museum of Civilization)

The four rectangular patches of vegetation mark the remains of Dorset culture houses on the south coast of Ellesmere Island. A small community probably lived here for a few weeks about 1,000 years ago. (Robert McGhee, Canadian Museum of Civilization)

This ivory image of a polar bear (138 mm long) was carved by an artist of the Dorset culture in Arctic Canada approximately 2,000 years ago. The lines and crosses on the surface of the body represent the bones and joints of the skeleton. This skeletal view, combined with the "floating" posture that is not characteristic of a live bear, suggests that the image may represent a spirit-creature. (Richard Garner, Canadian Museum of Civilization)

The colours of a Siberian town: the "Hotel North" in Lavrentia. (Robert McGhee, Canadian Museum of Civilization)

Children visit a newly caught grey whale while hunters dispose of the head of their earlier catch. (Robert McGhee, Canadian Museum of Civilization)

Thingvellir, the site of the first Icelandic parliament, remains a locality of great emotional significance in contemporary Iceland. The original parliament was held in the amphitheatre formed by the rifted basalt ridge in the foreground, with the lawspeaker presiding from a high rock in the wall at left. Participants in the meeting camped on the plain below the ridge, which provided water and grazing for their horses. (Ragnar Th. Sigurdsson)

Reconstruction of the small Viking-period house, the remains of which were excavated at Eiriksstadir in Haukadal, western Iceland. (Patricia Sutherland, Canadian Museum of Civilization)

The remains of Hvalsey church stand beside a fiord in southern Greenland. This mediaeval structure was the site of a wedding in 1403, which is described in the last historical record known from the Greenlandic Norse settlements. (David Keenlyside, Canadian Museum of Civilization)

John Ross and his crews encounter the Inughuit of northwestern Greenland in 1818. Drawn by Ross's Greenlandic interpreter John Sackheuse, this picture shows an interesting combination of European and traditional Inuit styles. (From John Ross, *A Voyage of Discovery…for the Purpose of Exploring Baffin's Bay*. London: John Murray, 1819)

The remains of an early Thule culture winter village on the gravel beaches of Brooman Point, High Arctic Canada. Collapsed whale bone roof supports fill some of the houses, and the village is surrounded by vegetation nourished by the refuse left here about 800 years ago. (National Air Photo Library, Ottawa, Canada)

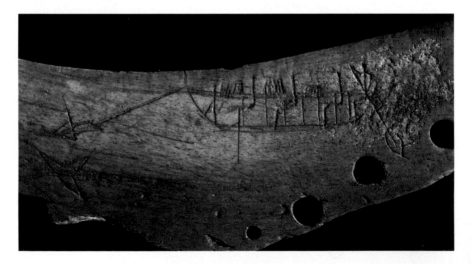

This knife handle, carved from whale bone and engraved with a scene of hunters in an umiak harpooning a large whale, was excavated from the Thule culture Inuit village near Qausuittuq. These early whalers had a much different set of hunting skills than the people resettled here during the 1950s. (Robert McGhee, Canadian Museum of Civilization)

receive the official attention of a major state organization. In the romantic ideology of Marxist communism, the reindeer herders, fishers and hunters of the north were "primitive communists" who worked and held all of their possessions communally. This was a grave misunderstanding of the actual situation, but although "primitive communism" had to be abolished before socialism was established, it was considered less offensive than capitalism or other systems of exploitation. The diverse mosaic of northern tribes was classified into twenty-six official nationalities, and became known as the "Small Peoples of the North." A powerful and enlightened Committee of the North was established in 1924 to oversee northern developments, and in the fervent wooden prose beloved of Soviet administrators, the committee was soon able to report that

> *The three ploughshares of Soviet power have been put into action: self-government, cooperation and enlightenment have already ploughed the virgin soil of the taiga and tundra, and have called forth from social and national non-existence a whole number of peoples hitherto almost unknown to the world, who have already started to join the ranks of fighters for socialism.*

The establishment of schools, local councils and a network of cooperative economic enterprises transformed "primitive communism" into contemporary socialism. Northern peoples had always carried out most activities in a cooperative fashion, and it was usually no great burden to give traditional family-based hunting parties or herding crews a new name and thus transform them into socialist cooperatives. Local councils had little authority, and were usually run by their Russian secretaries, but provided a forum to air grievances. Siberians who had previously been defenceless before the abuses of distant and powerful kingdoms now found themselves part of a new, paternalistic but generally benevolent order.

This situation lasted for about a decade, before the Stalinist forces of rural collectivization began to penetrate the taiga and eventually the tundra. The Committee of the North was abolished in 1935, and several of its members executed. The administration of some regions was turned over to the state company charged with the industrialization of Siberia using prison and exile labour; Arctic regions and their peoples became the responsibility of the Directorate of the Northern Sea Route. The heads of successful reindeer-herding families were now redefined as *kulaks* (rich peasants who exploited their neighbours) to be isolated, exiled or even executed and their herds confiscated to form the basis of state-owned collective farms. As elsewhere in rural Russia, these measures were met with local defiance and occasionally with armed resistance. Massive numbers of reindeer were hidden or slaughtered—perhaps one-third of the total Siberian stock. For their owners, the new laws meant not only the disappearance of personal wealth and influence, but the violation of cultural and religious taboos prohibiting them from transferring ownership of the individual animals to which their lives were so intimately tied. Some isolated groups drifted off with their reindeer, taking refuge in distant and inaccessible regions of the tundra and taiga. Others complied in name only, maintaining subterfuges that were successful as long as the attention and energies of the Soviet state were otherwise engaged, first by the purges emanating from Moscow during the 1930s, and then by the Great Patriotic War of 1941–45.

By the 1950s Siberia was ripe for the most radical period of transformation that had ever been experienced by the people of the region. Collectivization was now an accomplished fact, and tribal or ethnic independence a thing of ancient stories. Transportation and communication technologies perfected during World War II opened the tundra to industrial ventures centred on mining and the development of oilfields, as well as to military enterprises associated with the new Cold War. The sudden intrusion of southerners, who brought with them new technologies and new economic and administrative interests, imposed an entirely different context on the lives of northern peoples. Siberian hunters and reindeer-herders who had long benefited from trade relations with outsiders, and in recent decades had made at least a pretence of accommodating their lives to the organizing principles of state socialism, now suddenly found themselves in a world in which they were neither sovereign nor of much interest to their new rulers.

Like their indigenous compatriots in northern North America, who faced the same onslaught of southern interest and authority during the mid-twentieth century, the earlier freedom of northern Siberians to vanish into the tundra came to an end. Many nomadic peoples were settled in permanent communities in the interests of providing education and medical services, as well as access to jobs in new industrial ventures. The removal of children to state-run boarding schools added another incentive to abandon the seasonal migrations that had been part of many groups' yearly rounds for centuries.

The end result of this process—which we will look at more closely in a later chapter—was to transform northern peoples from a mosaic of independent societies, each with a cherished language and culture of its own, into a relatively homogenized population of town-dwellers whose culture, language and social allegiances were increasingly Russian. The land-based skills of hunters and herders were devalued by formal education, and by employment in the unskilled and semi-skilled jobs that maintained northern towns and industries. Hunting and herding were transformed into industrial operations organized on the model of *kolkhoz* (communal farm) and later *sovkhoz* (state farm) enterprises in the agricultural regions of the Soviet Union. Hunters and herders, together with mechanics and cooks and bookkeepers, all became salaried employees of the state. Veiled suggestions that indigenous peoples should be compensated for the use of their land by the state, or that local cultures should be encouraged to exist, were buried beneath the overwhelming sense that local peoples had been absorbed into a far greater and more advanced society—a society that provided the hitherto unknown benefits of education, heated apartments, medical attention, television, regular wages and pensions, even occasional vacations at Black Sea resorts.

And then, after two generations of Siberians had abandoned their traditional cultures to participate fully in the Soviet system, the system itself dissolved. During the early 1990s, wages and pensions began to arrive late, or not at all. Markets for local produce disappeared. Spare parts for vehicles and other equipment were unobtainable, and even food and fuel supplies became scarce and unreliable as transportation systems deteriorated. The centralized authority no longer existed to provide for the economic needs of communities at the distant margins of the country, nor could the economy afford to support these communities as it had in the past. Many of the Russians and other outsiders who had been attracted to the north by generous wages and living allowances now left the region, returning to central Russia where economic conditions were not as drastically affected by the events of the 1990s.

Northern Siberians didn't have the option of retreating to a more prosperous homeland, but at least they now had the right to ask for the return of their own land, destitute and mutilated though it was by a half-century of failed industrialization. Organizations calling for the rights of indigenous peoples, for aboriginal land claims, for the revival and promotion of local cultures, had begun to be heard in the late 1980s during the years of *glasnost*, an official policy of open discussion. With the economic and social upheaval of the 1990s, and the withdrawal of centralized economic support, these groups became increasingly significant voices in Siberian affairs. The need for vocal native organizations was in large part a reaction to the increased commercial opportunities available in the north to non-natives, entrepreneurs whose actions were no longer kept in check by the restrictions of a paternalistic central government. The forestry, mineral and hydrocarbon resources of Siberia were now open to exploitation by both foreign corporations and local ventures, which operated with scant concern for environmental protection or the needs and wishes of local communities.

The calls for indigenous title or control over traditional land and its resources have not yet had much result. As part of the move to privatize the Russian economy, some reindeer herds have been turned over to family-based groups of herders. Some herding ranges and hunting lands have been obtained by similar groups on long-term lease-like contracts that fall short of privatization of the land. Most northerners, however, find little advantage in these or other arrangements made possible by the post-socialist economy. Most long for the security, assured salaries and other benefits provided by socialism. The Brezhnev period, when the economy of the Soviet Union was in the throes of its final illness, were golden years for the people on the margins, whose way of life could not be supported by any rational economic system. The collapse of the Soviet Union has left the peoples of northern Siberia bereft of both their traditional way of life and of the socialism that supplanted it.

This was the Siberia that I encountered when I arrived on a summer morning in the Chukotkan town of Lavrentia. My first impressions did not allay the sinister preconceptions and fears that I carried with me from the West. A gusting wind carried icy fog from the Bering Sea, mixed with the sulphurous coal smoke streaming from several black metal stacks. Decrepit-looking trucks lurched down a mud-rutted street between unpainted anonymous buildings. Across the airstrip an unmanned guard-tower overlooked a range of abandoned barracks, a sight that was disturbingly reminiscent of films depicting the concentration camps of mid-twentieth-century Europe. The few people on the streets were drably dressed, carrying meagre shopping-bags or walking with no apparent purpose, their faces determined or indifferent. A brown-green vehicle resembling a military armoured personnel-carrier roared down the main street, mud flying from its clattering tracks. From the edge of the town, treeless hills rose to a low ridge capped with huge military radar screens and an array of satellite dishes. The jetlag I was suffering after the long flight from Moscow, together with the disorientation of culture-shock, provided the distorting lens through which I saw in this small town all that I had previously supposed Siberia to be, from Solzhenitsyn's Gulag Archipelago to Marco Polo's Land of Mud and Ice.

However, after a few hours of sleep in the small hotel, whose staff worked with surprising efficiency and dedication, I awoke to a transformed world. Glittering sunlight had washed the drabness from the town. To Western eyes, the first impression of a 1990s Russian community is the absence of colour. Without neon signs, billboard advertising or the hard brilliance of the painted metal and colour-impregnated plastics that fill our streets, a community of wood and stucco and official vehicles appear uniformly drab. But gradually the eye begins to see the colours of the town. Chalky blues and greens glow gently from the window frames and wooden trim of the older buildings. Washes of deep red ochre decorate stucco walls, a painted reindeer-sled drives into a muted sunset across the front wall of the hotel, and the kindergarten bears a massive rainbow of soft yellows, reds and blues.

The edge of town was marked by a small stream, across which I balanced on a driftwood log to find myself on a beach backed by a line of wooden huts. An elegant and freshly painted wooden whale-boat, built to the traditional pattern used by the nineteenth-century American whalers described by Herman Melville, was drawn up on the gravel. An identical boat lay bow-on to the beach, moored to the torpedo-shaped corpse of a whale surrounded by a cloud of red water. In the open front of one of the huts, the successful hunters drank tea around a small woodstove. The news of a successful hunt had spread through the town, and children and adults were gathering to view the whale. A crocodile of kindergarten children carefully negotiated the stream across the beach; kids with Asiatic faces and black braids holding hands with tow-haired Russians, all freshly washed and dressed in bright wool, they formed a chattering cluster around the dead animal. Most people chewed small pieces of whale skin that an old man in hip-boots sliced from a square hole in the animal's flank.

Eventually the hunters finished their tea and a tractor arrived to tow the animal ashore, but first the skull and cleaned bones of the last whale captured were dragged from the beach and deposited in deep water. This may have been done for sanitary reasons, but among most northern peoples the proper disposal of animal bones is an important aspect of living properly, and the action had elements of an old ritual designed to prevent offence to the whales. The work was supervised by an energetic middle-aged man whom the others called Volodya; he had apparently led the hunt and now was taking charge of the disposal of the bones and the distribution of the meat. I don't know what combination of traditional sharing and commercial transaction governed the distribution, but everyone on the beach seemed to get a share, and whale meat *kutlety* appeared on the hotel dinner menu that evening. In the new post-socialist Siberia, a collapsing economy has encouraged the re-emergence of the flexible

The tracked vehicle called *vesdekhod*, such as this one on the street of Lavrentia in Chukotka, is the workhorse of the Siberian tundra. (Robert McGhee, Canadian Museum of Civilization)

systems of cooperation and sharing that had served indigenous peoples for many centuries before 1917.

A ten-minute climb over the hillside behind the whalers' beach brought me to a different world in which I was suddenly at home. The soot and road dust, the dirty smokestacks and the rumbling trucks were suddenly gone behind the shoulder of the hill, and the scene was empty of humans and the litter of human technology. All around were green slopes swelling to a distant horizon, the sparkling blue of the bay, and distant mountains streaked with snow. My feet followed a narrow path of peaty earth winding through an ankle-deep mat of herbs and flowers, as brilliant and complexly patterned as any Persian carpet. Small pillows of yellow and purple saxifrage were set amid fields of golden Arctic poppies, showy louseworts and pink willowherbs; flocks of shaggy white cottongrass flowers nodded in the wind along the edge of a bubbling stream; willow trees with twisted trunks the size of pencils supported huge catkins, and tiny prostrate birches hugged the earth for warmth. A flock of snow buntings flitted from one lichen-covered boulder to the next, and a huge bumblebee droned across the path. The sun-warmed tundra was an immense and very familiar garden, a garden that extended for thousands of kilometres westward to Arctic Europe and eastward across the Bering Strait to Alaska and Canada where I had first learned to name its flowers and recognize its birds.

Over the following weeks I found that much more than the tundra of Chukotka was familiar. As elsewhere in Russia, the indifference or hostility of officials was dissolved by the warmth, helpfulness and generosity of local people. These were individuals whose problems and concerns were similar to those of people in other Arctic regions: providing for one's family; dealing with the outsiders who disparaged local culture and values while assuming they had the right to use local resources; trying to understand why alcohol, drugs, disease and violent death plagued one's community; and searching for a way to combine the benefits of global culture and economics with a secure sense of belonging to one's own particular place.

Several weeks later, when the first showers of snow heralded the end of summer weather, we left the archaeological dig where I had been working and began the long trip home. The first leg was by *vesdekhod,* chartered from the remnant of the local state farm. This was the tank-like vehicle that I had first seen roaring down the main street of Lavrentia; I now knew that these unstoppable machines were the workhorses and communication links of the reindeer-herding collectives and other communities that lived beyond roads. From earlier experiences with these vehicles we had learned that the roof was more comfortable than the dark and cramped interior, and our drivers had assured us that we might see bears on this trip. In a foggy evening twilight we clattered and splashed for several kilometres up a wide gravel riverbed, then ground through darkness and snow along steep and narrow valleys illuminated only by a spotlight on the roof. We saw no bears, only occasional fuel drums left by previous travellers, and sleep was impossible: even without the continuous threat of being thrown from the pitching roof and left behind in the tundra, the jolting and exhaust fumes would have kept us awake.

Eventually the track began to descend, and in the midnight darkness we clattered into the abandoned coastal settlement of Pinakul, a wasteland of rusted barges, decrepit machinery, broken vehicles and collapsing buildings. Here we were to spend the night and continue the following morning by boat. The drivers disappeared into the darkness to waken the only inhabitants of the derelict settlement, and we recognized our host as the same Volodya who had led the whale hunt in Lavrentia several weeks before. Vladimir Eivoucheyvoun is a Chukchi hunter who turned his back on the alcohol and despair of town life, and set out to make a fresh start in the new world of the 1990s. Through the sleet and darkness and corroded debris he led us to the only occupied house, a tall eccentric structure that he had built himself. In a warm room surrounded by sleeping children his wife served tea boiled on a pressure-stove, bread and jam and pickled salmon. Volodya led us up wooden ladders to a glass-paned greenhouse built onto the roof of this unusual and hospitable home, and we settled to a few hours sleep on beds of caribou skins in a room full of cucumber vines. At dawn we were wakened for the final part of the journey, a crossing of the bay to Lavrentia in a *baydar* built and skippered by Volodya's father. This traditional Chukchi boat, covered with walrus hide tied over a wooden frame, was driven by a powerful Japanese outboard motor and was undoubtedly capable of carrying whale hunters far out to sea, or of crossing Bering Strait to Alaska. At the end of my visit to Chukotka I had found myself among a family that had chosen a unique and rewarding mix of the old and new ways, and were surviving well in the changing world that threatened so many around them.

5 VIKINGS AND ARCTIC FARMERS: THE NORSE ATLANTIC SAGA

ON A WARM AND CLOUDLESS AUGUST AFTERNOON the Viking ship *Gaia* approached the coast of North America. The huge sail pumped gently with a steady following breeze as the ship rode, lightly as a resting gull, across the ridges and valleys of a long north-easterly swell. Sunlight glinted from icebergs and flared on the distant spouts of whales. The grey cliffs of Newfoundland rose to starboard, and the foaming wake curled away towards Greenland. This tiny, fragile shell of wood had steered through fields of deadly ice, ridden over shoals that would have torn the bottom from a deeper ship, and passed through storms while taking no more than spray over the rails. Its crew was as confident and comfortable as if they were aboard a modern sailing yacht.

Gaia is a replica of the Gokstad ship excavated from a Viking burial mound near Oslo, and in August 1991 it was completing a crossing of the Atlantic from Norway by way of Iceland and Greenland. I had joined the ship a few days earlier, and was now entrusted with the rudder for the first time while the rest of the crew busied themselves with preparations for going ashore. As I anxiously tried to keep the wake as straight as those drawn by the professional sailors aboard, my mind kept veering into the past, to identical frail

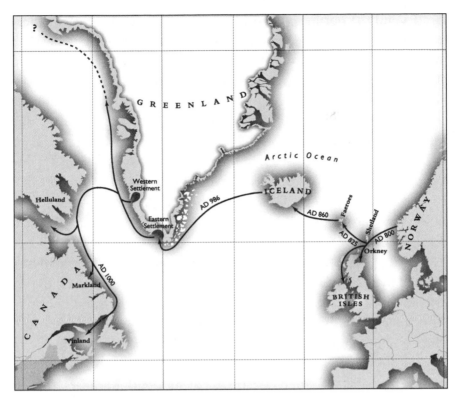

Viking Expansion Across the North Atlantic

vessels that had made similar landfalls along this coast a thousand years ago. A millennium of history rode with *Gaia* that afternoon, the history of early exploration and settlement across the island countries of the North Atlantic, and the unique Arctic venture of a small medieval society that flourished for almost 500 years.

The Viking Age began about AD 800, at a time when Europe was beginning to emerge from the period that would later be known as the Dark Ages. Charlemagne's Holy Roman Empire now stretched from the Pope's capital of Rome to the coasts of the North Sea, from the Muslim kingdoms of Spain to the margins of the eastern European plain. After four centuries of barbarian invasion and local conflict the Roman Empire had reappeared in a new form. The spiritual power of the Church now combined with the temporal strength of anointed Christian rulers to create the medieval civilization that was to flourish in Europe for the next half-millennium.

Beyond the northern boundaries of this Europe lay a very different world, one that had experienced neither invasion by the armies of ancient Rome nor the mixed blessings of Roman civilization. These were lands of constantly brawling warlords, men whose need for the constant display of wealth and personal power stimulated artistry, craftsmanship and foreign ventures in search of costly, rare and exotic objects. Some of these societies—the Irish and the Picts of Scotland—were Christian, but those of the North Sea and Baltic coasts were the last Europeans to worship the pagan pantheon that had once ruled Europe. These were the societies that produced the Viking raiders who began to appear off the shores of Christian Europe in the years around AD 800.

To the people whose churches and monasteries were looted and burned by these heathen marauders, the northern pagans exemplified the forces of evil and mindless brutality. This verdict wasn't entirely just, since destruction and casual violence were common enough throughout ninth-century Europe. The Vikings were unique only because—unlike Christian raiders—they didn't stop at the church doors. Their victims were also mistaken in seeing the marauders as the representatives of an entirely violent society. The raiders whom they encountered were merely the sharp edge of that society—the young men sent off each summer to gather the wealth and slaves required to validate the positions of their masters in the development of a flourishing northern civilization. Behind them lay communities of craftsmen, artists, merchants and, most importantly, farmers.

The Viking eruption into Europe was fed by social needs, but it was made possible by an advance in military technology: the development of ships that allowed the peoples of Scandinavia to undertake long sea-crossings and to appear with disconcerting suddenness on distant coasts. The finest example of such a ship was excavated from a burial mound at Gokstad near Oslo in 1880. The Gokstad ship was built in the ninth century, and at the end of its useful life was used as the burial-vessel of a local ruler whose skeleton—along with those of his horses, dogs and a peacock—was found aboard. The ship was built of thin overlapping oak planks fastened to one another by iron rivets, the entire shell-like hull tied by cords to a light wooden frame.

Gaia, a reconstruction of the Viking-age Gokstad ship, approaches the Newfoundland coast during the summer of 1991. (Robert McGhee, Canadian Museum of Civilization)

The single mid-ship mast would have supported a large square-sail, ports for sixteen oars line either side of the hull, and a rudder is attached to the starboard side. The hull is twenty-four metres long, five metres wide, and would have drawn less than a metre of water when fully loaded with a crew of about thirty-five men. Replicas of the Gokstad ship have demonstrated that it could travel at speeds of over twelve knots, and was capable of crossing not only the North

Sea but the North Atlantic in relative security. The *Gaia* that I steered in 1991 was a replica of the Gokstad ship.

The forces that propelled Viking raids against Christian Europe also drove a constant quest for new uninhabited farmlands where ambitious men might establish themselves in the manner practised by local chieftains in their fully occupied home countries. Norse ships were highly seaworthy, but sailors had no compasses or any other adequate means of navigating in the open sea, and the vessels were frequently driven before storms for days on end. This inability to follow a planned course in stormy weather led to the accidental discovery of many islands. When ships wandered far in aimless attempts to regain their bearings and their home ports, it was inevitable that all of the mountainous islands of the North Atlantic would soon be discovered.

The first of the stepping-stones that eventually led Viking sailors as far as Greenland and Arctic Canada was probably reached during the eighth century as part of the movement that took raiders to the coasts of Britain and Ireland. The Hebrides and Orkney islands, lying only a few kilometres off the Scottish coast, had long been settled by Irish and Pictish peoples, as had Shetland almost 100 kilometres further north and midway between Scotland and Norway. From here it is over 250 kilometres northwesterly to the Faeroes, which lie midway between Scotland and Iceland, and information on the existence of these steep islands was probably obtained by the Vikings during their early ventures to Ireland.

To any observant inhabitant of Ireland, Scotland or the Hebrides, the existence of distant lands in the far north must have been apparent. Each spring vast flocks of swans, geese and lesser birds arrived from the south, spent days or weeks feeding frantically, and then in increasing noise and excitement lifted off and gained altitude until they disappeared over the empty sea to the northwest. In the autumn, weary and bedraggled flocks reappeared with youngsters newly fledged. The Vikings were not the first Europeans to follow these guides into the unknown realms of the North Atlantic. This distinction goes to the Irish, whose explorations were driven not by the search for wealth and personal renown, but for distant hermitages where monks could live their lives in worship, protected from the temptations of society by the fastness of the sea. For centuries Irish hermits had launched themselves in skin-covered currachs driven by oars and square-sail, trusting in God to preserve and direct them on their holy missions.

We do not know how far these explorations took them, but intriguing hints of distant voyages survive. The *Navigatio Sancti Brendani* is an implausible story written in the ninth century, telling of a journey taken in AD 565–572 by the elderly Saint Brendan in search of an earthly paradise. Among other wonders, the saintly crew encountered a mountain that belched fire and threw out glowing coals, a description suggesting volcanic Iceland. Brendan's report of the wind suddenly dropping and the sea becoming calm and coagulated, an apparent description of sea-ice forming, indicates that they were probably far to the north, as does their encounter with a crystal column rising from the sea, the colour of silver and harder than marble, surely an iceberg. North Atlantic icebergs calve from the glaciers of Greenland and drift south with the Labrador current to eventually melt in the warm waters of the Gulf Stream; they rarely drift to the east of 36° West longitude, which would place this apparent sighting in the mid-Atlantic, approximately 2,000 kilometres to the west of Ireland. This fantastic tale seems to suggest that by the ninth century, when it was written, the Irish knew of Iceland and may occasionally have voyaged even farther to the north and west.

A far more reliable source on early Irish voyaging comes from a book written in AD 825 by Dicuil, an Irish monk and geographer living at the court of Charlemagne. In his book *Liber de mensura orbis terræ*, Dicuil reports that

> *There are many islands in the ocean to the north of Britain which can be reached from the northernmost British Isles in two days' and nights' direct sailing with full sails and an undropping fair wind. A certain holy man informed me that in two summer days and the night between, sailing in a little boat of two thwarts, he came to land on one of them. Some of these islands are very small; nearly all of them are separated one from the other by narrow sounds. On these islands hermits who*

have sailed from our Scotia [another name for Ireland] *have lived for roughly a hundred years. But, even as they have been constantly uninhabited since the world's beginning, so now, because of Norse pirates, they are empty of anchorites, but full of innumerable sheep and a great many different kinds of seafowl. I have never found these islands mentioned in the books of scholars.*

This description must refer to the Faeroes, the first major stepping stone across the far northern Atlantic, and indeed archaeological evidence suggests that Norse farmers arrived in these islands at some time around AD 800, shortly before Dicuil's book was written. The learned monk also tells of an island named Thule that lies farther to the north:

It is now thirty years since priests who lived in that island from the first day of February to the first day of August told me that not only at the summer solstice, but in the days on either side of it, the setting sun hides itself at the evening hour as if behind a little hill, so that no darkness occurs during that brief period of time, but whatever task a man wishes to perform, even to picking the lice out of his shirt, he can manage it precisely as in broad daylight.... They deal in fallacies who have written that the sea around the island is frozen, and that there is continuous day without night from the vernal to the autumnal equinox, and vice versa, perpetual night from the autumnal equinox to the vernal; for those sailing at an expected time of great cold have made their way thereto, and dwelling on the island enjoyed always alternate night and day save at the time of the solstice. But after one day's sailing from there to the north they found the frozen sea.

As I noted in Chapter 3, the name Thule comes from classical geography and refers to a land located six days' sail north of Britain that was visited by Pytheas in the fourth century BC. However, the island described by Dicuil seems almost certainly to be Iceland, situated just below the Arctic Circle where the midsummer sun barely sets, and where sea ice frequently lies just off the northern coast. The existence of Irish hermits on Iceland is confirmed by both Norse historical accounts and by place-names derived from *papar*, the Norse name for these holy fathers. Shortly after the time that Dicuil was writing, the monks of Iceland, like those of the Faeroes, found their hermitage invaded by Norse farmers hungry for land. Earlier I suggested that the Irish probably found their way to these far northern countries by noting and then following the paths of migrating geese, swans and other fowl. The Norse, in their turn, had the migratory paths of Irish hermits to guide them to new lands.

Icelandic tradition places the Norse discovery of the island around AD 860 and tells of three early venturers: the Swedish Viking Gardar Svavarsson, who circumnavigated the island and named it Gardarsholm; Naddod, who named it Snowland when a snowstorm covered the mountains as he sailed away; and Floki Vilgerdarson, who navigated with ravens as Noah had with doves, and bestowed the name by which the country has since been known. Although these explorers returned with mixed reports on the quality of the land and climate, they agreed that Iceland was a large country. The unanticipated size of the place—the scale of the mountains and the rivers, the breadth of the huge green valleys leading to the far interior—is just as impressive today. To the traveller crossing it by aircraft or navigating a rented car around the coastal highway, Iceland seems an isolated continent rather than a small island.

The Iceland seen by the first explorers was considerably more verdant than it is at present, after its vegetation has been stripped by a millennium of tree-cutting and sheep-farming. According to *Íslendingabók*, the twelfth-century Book of the Icelanders, "At that time Iceland was covered with forest between mountain and seashore." The forest was composed of birch trees, the same small contorted trees that today grow in locations—such as country graveyards—where they are protected from sheep. At that time the northern hemisphere was entering a phase that has become known as the Medieval Warm Period, a climatic episode when summers were significantly longer and warmer than those of recent times, and sea ice was less pervasive than at present. Farmers from Norway, and from the Norse settlements in the northern British Isles, would have recognized good grazing land from the stories that early explorers brought back from Iceland. The tales must also have

noted rivers swarming with trout and salmon; inestimable flocks of
nesting geese, swans and seabirds; seals and whales and the walrus
that would provide not only meat but a fortune in ivory. The attraction was so great that King Harald Fairhair—whose taxes, tyranny
and restrictions on rival chieftains had a significant effect on
encouraging emigration—became concerned that Norway would
be depopulated, and placed a heavy tax on would-be emigrants. By
the end of Harald's reign, about AD 930, Iceland had a population
estimated between 10,000 and 30,000 people, and all the arable
land had been taken.

The settlement of Iceland would seem to have been a recipe for
human disaster. The country lay isolated on the fringes of the
unknown Arctic, the environment marginal and poorly understood.
The settlers who filled the land were the most ambitious and impa-
tient produced by Viking society, and soon began to quarrel over
their privileges, slights to their honour and the boundaries of their
farms. Slaves made up a significant proportion of the population,
many of them Irish warriors captured in battle, who were as arrogant
and freedom-loving as their Norse owners and as easily roused to
murder and revolt. There was no law, and no means of enforcing any
law. Nor could religion serve as a unifying force, since some settlers
were Christians from Ireland and the Hebrides while most retained
allegiance to pagan gods. A sociologist could not be faulted for pre-
dicting that such a situation would spiral inevitably into anarchy and
disaster. Yet from such beginnings, the Icelanders built a society that
has lasted over a thousand years and is noted for its stability, its egalitarian regard for the community, and its respect and devotion to
history and tradition.

The road was neither straightforward nor inevitable, but it is at
least well known since Iceland also became the first truly literate
country of northern Europe. The importance of literature and history in Icelandic culture is often credited to the Irish and Hebridean
component of the early Icelandic population. Certainly no other
Norse society developed a literary tradition on the scale of the
Icelandic sagas, nor the interest in historical recording that preserved
a unique record of an early European society. Soon after the Roman

alphabet was introduced by the Church, it was adapted to the Icelandic (Old Norse) language and spread throughout the country during the twelfth and thirteenth centuries. Most of what we know of the pagan religion of northern Europe comes from the *Eddas,* written by thirteenth-century Icelandic scholars. Over one hundred family sagas, many preserved as oral tradition over the centuries before being committed to writing, paint vivid pictures of a brawl-ing society fuelled by the quest for wealth, honour and revenge. Churchmen and farmers compiled annals recording events of local interest and news obtained from travellers who had visited distant parts of the Norse world. Careful historians reconstructed lists of the families who had participated in the original land-taking, together with the names of their farms. They described the development of a system of laws about AD 930, based on an annual council of the leading members of society—the *Althing*—which has been por-trayed as the first European parliament.

The annual Althing was held at Thingvellir in southwestern Iceland, in a geologically astonishing natural auditorium, with an adjacent plain for camping and the pasture of horses. Thingvellir has been preserved in something like its original condition, and remains the symbolic centre of the Icelandic state. On a summer day one can stand in the assembly-place, a grassy vale surrounded by vertical walls of black rock torn apart by vast geological forces, and sense the his-torical significance that has accumulated over a thousand years of use. It was at this time of year that people from around the country trav-elled the horse-trails to Thingvellir and set up their tents and turf-walled booths on the plain below the cliffs.

The Althing was an important social and economic occasion for the Icelandic people, but like its rival Mother of Parliaments at Westminster, the Althing had little to do with democracy. Power was held by thirty-six *godar*, men with sufficient wealth and influence to support temples and to be the effective religious and secular chief-tains of local regions. The manoeuvrings and power plays of this wealthy elite were always the backdrop to cases of law considered by the Althing, yet without this structure it is difficult to imagine how anarchy could have been avoided. The effectiveness of the system was

This tumbling waterfall in northern Iceland is named Godafoss (Waterfall of the Gods), from the heathen idols which were thrown into it when Iceland officially adopted the Christian religion in AD 1000. (Robert McGhee, Canadian Museum of Civilization)

well demonstrated in AD 1000, when Christianity was advancing through the population and a potentially perilous decision had to be taken regarding the religious future of Icelandic society. After a debate during which zealots from both camps promised civil war, the question was given to Thorgeir the Lawspeaker or president of the parliament. Thorgeir retired for a day and a night, and then presented a decision that was to prove both practical and wise: that all Icelanders would be baptized as Christians, but that certain of the old heathen laws would still stand; people would still be permitted to eat horse-meat; and those who wished to practise the old religion in private could do so. It is reported that many were baptized in the hot springs of Iceland's volcanic landscape, since their zeal for the new religion did not extend to cold water baths.

The Icelandic Republic lasted for three centuries (a century longer than any republic of the modern world) before losing its political independence to the Norwegian king in 1262. It would not regain its autonomy until the Second World War.

A turbulent period led up to this event, driven by increasing competition in a situation of decreasing resources. By the thirteenth century the destruction of Iceland's forests—for firewood, as fuel for smelting iron and to provide grazing land—had begun to take its toll on the environment. Sheep and goats removed the remaining vegetation cover, the light volcanic soil eroded and drifted in the winds, and entire regions became barren wastelands. A cooling climate brought bitter winters, freezing livestock in the fields or starving them for lack of fodder. Sea ice began to threaten navigation, and commerce with Europe fell off as the increasingly scarce and irreplaceable Icelandic ships became derelict or were lost at sea. Civil wars broke out, and the smoke of burning farmsteads marked the progress of local armies down the broad valleys.

Allegiance to the Norwegian king brought a measure of peace to Iceland, but the following century was a time of even greater hardship for most of the population. Farms continued to be abandoned as the climate deteriorated and erosion worsened. Volcanic eruptions destroyed entire districts. Epidemic diseases—smallpox, plague and others less identifiable—that had reached Europe as a result of the Crusades and the opening of Asiatic trade, finally appeared in Iceland and decimated the population. In the early 1400s coastal farms began to be raided by English fishermen and pirates, a development that was eventually to change the fortune of the country. What attracted the English to these coasts was not the small items of value that might be looted or the young people who could be kidnapped and sold into slavery—although this was frequently done—but the codfish that swarmed in Icelandic waters. The Icelandic economy soon became tied to the cod trade, and in turning from subsistence agriculture to commercial fishing the Icelanders found an economic niche in the world that their country still occupies today.

Other world events were to play unexpected roles in Iceland's history. Most important was the Second World War, when Allied forces needed both a refuge for North Atlantic convoys and a refuelling stop

for aircraft being ferried from North America to the battle zones of
Europe. The American bases established at Reykjavik and Keflavik
brought Iceland into the heart of the Allied war effort and had a huge
effect on the local economy. The American materials and cultural
influences that were introduced in those years would have an ongoing
impact on the Icelandic way of life. And with Norway and Denmark
under German occupation, the war provided the opportunity for
Iceland to declare itself an independent republic, which it did in 1944.

Iceland holds a remarkable fascination for the historically minded.
This resolutely modern nation has such a distinctive past, and is so
closely built on and devoted to that past, that history is apparent
everywhere. Place-names are a constant reminder of historical events,
from the Westman Islands off the southern coast—named for the role
that they played as a refuge for rebelling Irish slaves—to Godafoss,
the turbulent waterfall on a northern river that was named for the
heathen idols that were thrown into the cataract when the country
became Christian. Farmhouses, looking much like their modern
contemporaries in northern Europe, are built on mounds of rock
and earth marking the remains of generations of turf-built farmhouses and cattle byres. Many of these farms can trace continuous
occupation from the Landnam (land-taking) period of the tenth
century. Driving Iceland's narrow highways, I have been occasionally startled by the appearance of a name familiar from the sagas on
one of the small green signs that identify farmsteads. The farm
named Bergthórshvoll on the south coast, for example, bears the
same name as that in which Njál and his sons were burned to death
by their enemies in one of the most tragic events in saga literature.
Archaeologists excavating at Bergthórshvoll have identified the
remains of a byre that was built during the late tenth century and
burned a few decades later, perhaps in the conflagration that
destroyed the adjacent house and its violent occupants.

In the pleasant valley of Haukadal in western Iceland, a small
Viking longhouse has recently been reconstructed from archaeological remains on a site next to a farm named Stora-Vatnshorn, which
provides bed-and-breakfast accommodation to visitors. According to
the sagas it was next to his father-in-law's farm at Vatnshorn that a

man named Eirik the Red cleared land and built his house after arriving in Iceland around AD 960. The local farmers of Haukadal during the past century still referred to the location next to Stora-Vatnshorn as Eiriksstadir (Eirik's place). Here, beside a tiny brook descending the steep sheep-grazed hillside towards the lake in the valley below, archaeologists excavated the remains of a tenth-century hall built of turf walls above a stone foundation. A thousand years of historical tradition may be correct in linking this structure to the saga-figure of Eirik the Red. In the words of Gudmundur Olafsson of the Icelandic National Museum, displaying the archaeologist's innate distrust of literary sources, "If there really was a man named Eirik the Red, this was probably his house." It is this saga-hero who was at the centre of the next Norse adventure into the western Atlantic, which eventually brought Norse farmers and hunters into the true Arctic and westwards as far as Canada.

While Iceland scarcely touches the Arctic Circle, Greenland is truly an arctic continent. Its southern tip extends barely to the six-tieth parallel, and the immense ice-cap that buries most of the land is a remnant of the great glaciers that covered much of the northern hemisphere during the last ice age. When European explorers began to visit the coast of Greenland during the sixteenth century they found the country occupied by Inuit, relatives of the aboriginal nation of hunters and fishers whose lands stretch across Arctic Canada to Alaska and Chukotka. Yet the long fiords and inland val-leys of western Greenland are scattered with hundreds of untidy mounds of turf and rock, the remains of farmhouses and byres that once protected small herds of cattle, sheep and goats. Occasional frag-ments of drystone walls emerge from the grassy landscape, marking the remains of churches in which no service has been held in more than five centuries. This narrow land, squeezed between an Arctic sea and the massive dome of the inland ice, was once the most distant outpost of medieval European civilization.

Given the uncertainties of Viking navigation, it is not surprising that land was soon discovered to the west of Iceland. The east coast of Greenland lies less than 600 kilometres from Iceland, less than three days' sail in good circumstances. Moreover, for much of that distance a sailor would be in sight of either the white volcanic cone of Snaefellsjokul (Snow Mountain), marking Iceland's western tip, or of the Greenlandic mountains, which rise to over 3,000 metres behind Ammassalik and came to be known to the Icelanders as Hvitserk (White Shirt). According to *Landnamabok* the discovery was made around AD 900 by a storm-tossed sailor named Gunnbjorn, and the new land became known as Gunnbjorn's Skerries. The area may have become a temporary refuge for Icelanders fleeing justice or revenge, but the one account of such an event preserved in *Landnamabok* tells of a hard winter during which the house was buried under snow, followed by a springtime of hunger and manslaughter.

If we can believe the sagas, the first successful attempt to survive in the new land occurred in AD 982 when Eirik the Red—whom we last saw clearing his farm in Haukadal—set out to find a refuge where he could spend three years of enforced exile. Eirik's stay at Haukadal had ended when his slaves caused a landslide that destroyed a neighbour's house; he had moved next to an island in adjacent Breidafiord, but had become involved in another quarrel that led to the killing of neighbours. Eirik was considered a troublemaker, even by the standards of the time and place, and when the local Thing (the council of leading landowners) sentenced him to exile Eirik outfitted a ship and sailed west. Forsaking the grim southeastern coast of Greenland, he turned southward with the pack ice that incessantly guards that shore, and rounded Cape Farewell. Here he entered a set of fiords much like those of Iceland, sheltered from the sea and lushly vegetated with grassy meadows and thickets of dwarf willow and birch. Caribou grazed the green hillsides, the unfished streams were rich in salmon, seals swam in the iceberg-studded fiords or rested ashore with no fear of human hunters. Unlike any of the other North Atlantic lands discovered by the Norse, Greenland was empty of human habitation.

During his three years of imposed exile Eirik made plans to settle his newly discovered country, and must have contemplated the irony of a despised outlaw becoming the respected first settler and first citizen of a new republic. With the instincts of a realtor he named the country Greenland "for he argued that men would be all

Brattahlid is the traditional site of Eirik the Red's farm in southwestern Greenland. The remains of a Norse farm building are in the foreground, while the recent reconstruction of the small church built "not too close to the farm" for his wife Thjodhild stands just over the brow of the hill. (Patricia Sutherland, Canadian Museum of Civilization)

the more drawn to go there if the land had an attractive name." When his term of outlawry ended he sailed back to Breidafiord in western Iceland and began recruiting emigrants. The region had been devastated by famine during the previous decade, and Eirik's proposition soon attracted a sizeable following. In 985 a flotilla of twenty-five ships set out for Greenland, and the sagas report that only fourteen completed the journey; the rest were lost or turned back by stormy weather and huge waves.

We can picture the small fleet that straggled into the fiords of southwestern Greenland. The ships were variations on the type known as a *knorr*, the workhorse of the North Atlantic during Norse times, which we know from a well-preserved example that has been excavated from the bottom of Roskilde Fiord in Denmark. At about fifteen metres in length the *knorr* was shorter than the earlier Gokstad ship, but beamier and with a deeper hull that made it more seaworthy and capable of carrying larger cargoes. These were undecked ships, open to wind, spray and slopping wave-crests, but modern replicas have shown them capable (with luck) of standing up to dreadful North Atlantic storms. Such a ship was capable of carrying the family and slaves of a small farm, together with the tools and household goods that they needed to start a new life, a farm's complement of sheep and goats and dogs, a pair of Icelandic horses, and a few small and dreadfully seasick cows. After days or weeks amid the choppy seas and terrifying pack-ice of the far North Atlantic, the calm waters and warm breezes of a Greenlandic fiord must have seemed a paradise.

As the founder of the new country, Eirik selected the most promising land and built a farm at a place that he named Brattahlid in the far interior of Eiriksfiord. The remainder of the fleet fanned out to select the best farmland in nearby fiords, while some continued northwards along the coast to establish another set of farms in the deep fiords behind the present Greenlandic capital of Nuuk. Other immigrants followed in subsequent years, and by about AD 1000 most of the usable land was probably occupied. The smoke of burning scrub-forests rose over the fiords as land was cleared for hayfields and pasture, houses and byres were built in the Icelandic fashion from stone and cut turf, and the country was explored from the rocky islands of the outer coast to the margins of the inland ice.

Two other major events occupied the first generation of Greenlanders. The first was the sighting, by the crew of a storm-driven immigration ship, of a more extensive land to the west. This discovery led over the following decades to further exploration and the naming of three more lands: Helluland (Slab Land), a barren country that can probably be identified with Baffin Island and adjacent regions of Arctic Canada; Markland (Forest Land), a name that probably refers to the heavily wooded coast of central and southern Labrador; and Vinland (Wine Land), probably the coasts of the Gulf of St. Lawrence. Attempts to establish settlements in Vinland occurred about AD 1000 according to the saga accounts, but were quickly abandoned from fear of the native occupants of the region. Markland may have served over the following centuries as a source of timber for treeless Greenland, but we know practically nothing of how often it was visited. Nor is Helluland mentioned in many historical sources, although recent archaeological evidence suggests that it remained part of the Greenlandic world in subsequent centuries.

The second major event that occurred during the early years of the Greenland colony was the introduction of Christianity. *Eirik the Red's Saga*, one of the two saga accounts of the settlement of Greenland, credits the introduction to Eirik's son Leif, who was converted at the court of Norway's King Olaf Tryggvason. This seems a literary invention, but the timing of the event—coincident with the Icelandic conversion to the new religion in AD 1000—is probably accurate. The saga goes on to relate an engaging tale of the first church in Greenland, which was built for Eirik's wife, Thjodhild, in an episode that has received credible support from archaeology:

> *Eirik took coldly to the notion of abandoning his faith, but Thjodhild embraced it at once and had a church built, though not too close to the farm. This church was called Thjodhild's Church, and it was there that she offered up her prayers, along with those men who adopted Christianity, who were many. Thjodhild would not live with Eirik as man and wife once she had taken the faith, a circumstance which vexed him very much.*

In 1961 construction near Eirik's Brattahlid farm encountered human bones, and archaeologists who were called in found a Christian graveyard surrounding a tiny chapel built of turf. The church measures only six metres by three and could have held only a tiny congregation, but may well have been the first Christian church in the New World.

Within a century the Greenland colony had grown to perhaps 2,000 people, the majority clustered in the fiords of the Eastern Settlement near Brattahlid and a smaller number in the Western Settlement located in the complex of fiords around 64° latitude, 500 kilometres to the north. The archaeological remains of 250 farms have been identified in the Eastern Settlement area, and a further eighty in the Western Settlement, although they were probably not all occupied at the same time. As in Iceland, the economy was based on sheep and goats that were raised for their wool, meat and particularly their milk, which was stored as cheese or as skyr, the yogourt-like soured milk that remains a part of the Icelandic diet. It seems unlikely that any grain could be grown on Greenlandic farms, and the making of hay for winter fodder must have been the heaviest labour of the summer months. Cattle and even a few pigs were kept on the larger and more favoured farms, but could not have flourished in the Greenlandic environment. Hunting and fishing must have contributed a significant amount to the diet, and would have become increasingly important in later centuries when the climatic cooling and environmental degradation that impinged on Iceland also affected the Greenlandic fiords.

Despite its remoteness and isolation, Greenland was a resolutely European country and maintained the beliefs and practices of medieval society. The remains of seventeen churches have been identified in the Eastern Settlement, two or perhaps three in the Western. Most of these were small buildings associated with individual farms, but several were large masonry structures complete with bronze bells and stained glass windows, which served larger districts. A fourteenth-century inventory of the Eastern Settlement also reported an Augustinian monastery and a Benedictine convent. A bishop was first appointed to the country in 1124 at the request of a delegation of leading farmers—who presented the Norwegian king with a cargo of ivory, walrus hides and a live polar bear—but the early bishops seem to have resided in Iceland. The first bishop to actually live in Greenland arrived around 1210, and his seat was established at Gardar, the largest farm in Greenland. Situated on a lush strip of land at the narrow neck between Eiriksfiord and Einarsfiord, two of the major population areas in the Eastern Settlement, Gardar boasted a large stone cathedral measuring 32 by 16 metres and probably augmented by a separate bell tower. In the fertile hayfields around the cathedral archaeologists have identified a number of buildings including a large farmhouse that probably served as the bishop's residence, byres for more than a hundred cattle, and a system of irrigation channels leading from dammed ponds in the surrounding hills. For almost two centuries Gardar was the centre of power and wealth in Greenlandic society.

Greenland became a tributary of Norway in 1262, at the same time that Iceland recognized Norwegian sovereignty, but throughout its subsequent history the Church rather than any secular power was the important link through which the remote colony maintained its connection to European society. In order to preserve this link the Greenlanders paid tithes and other taxes in the goods their country produced: cloth woven from the fine wool of sheep raised at the edge of the glaciers, furs of arctic animals including polar bears, falcons for the sport of European and Middle Eastern kings, walrus hide cut into ropes for rigging ships and, most importantly, ivory from narwhal and walrus tusks. In 1327 a church official reported collecting Crusade tax in the amount of about 650 kilograms of ivory, which would represent the tusks of approximately two hundred walrus. It was the quest for these valued trade goods that took Greenlandic Norse hunters far to the north of their settlements, and probably brought them into contact with the aboriginal occupants of Arctic Canada.

The first indication of such contact comes from a twelfth-century text, the *Historia Norvegiae*, which states

> *Beyond Greenland, still farther to the north, hunters have come across people of small stature who are called Skraelings. When they are struck with a weapon their wounds turn white and they do not bleed, but if they are killed they bleed almost endlessly. They do not know the use of iron, but employ walrus tusks as missiles and sharpened stones in place of knives.*

This description would appear to fit the people known to archaeologists as Tuniit, the people of the Dorset culture who were the first aboriginal occupants of Arctic North America. (It has been suggested that the reference to bloodless wounds may have derived from a Norse skirmish with people who wore clothing sewn from animal skins, rather than the more permeable cloth familiar to the Norse.) The Tuniit at this time occupied the coast and islands of Arctic Canada, and the far northwestern corner of Greenland to the north of Melville Bay. Archaeologist Pat Sutherland has recently traced an intriguing trail of evidence suggesting that the Norse and the Tuniit occupants of the region known as Helluland (Baffin Island and adjacent areas) may have been partners in a complex and enduring relationship, perhaps centred on the trade in walrus ivory. The Tuniit were expert walrus hunters, and it would seem to have been in the Norse interest to obtain ivory by trade. It would be several decades before the Norse encountered a different and potentially more dangerous nation, the Inuit, who moved eastward and displaced the Tuniit at some time during the later twelfth or thirteenth century.

The history of Norse Greenland is much more poorly known than that of Iceland, and the eventual disappearance of the country's Norse society has been a perplexing mystery. Inadequate scraps of evidence have been assembled and refitted to support several theories, none of which are particularly convincing. The first intimation of trouble comes from the Western Settlement, the more northerly and more marginal cluster of farms. This hint is found in the report of Ivar Bardarson, a church official who was sent to Greenland about the middle of the fourteenth century to investigate the state of the Church and its property. Bardarson sailed to the Western Settlement, probably during the 1350s, and reported that "At present the Skraelings hold the entire Western Settlement. There are horses, goats, cattle, and sheep, but all wild, and no people, either Christian or heathen." The report survives only in late and recopied translations which have probably been corrupted in transmission, and this brief account has been endlessly debated, largely because it makes little sense. Farm animals, particularly cattle, could not have survived a Greenlandic winter at this latitude without human care, especially if the country was infested with Inuit hunters. Archaeological evidence suggests that Inuit groups may have penetrated southwards along the Greenland coast as far as the Western Settlements by the 1350s, but they would probably have limited their occupation to the outer coasts, and would have been more interested in trading with the Norse farmers who lived on the inner fiords rather than displacing them from what would have been relatively poor hunting grounds. It has been suggested that Bardarson exaggerated the extent of his survey, perhaps after encountering Inuit on the coast, or that the owners of the sheep and cattle may have simply been hiding from a person who would have been viewed as a tax collector.

Whatever the case, it is clear that the Western Settlement was in trouble, and it seems to have been largely or entirely abandoned at some time during the fourteenth century. Ice-cores drilled into the inland glacier have recently provided precise evidence on the past climate of Greenland. These show that the period from AD 1340 to 1360 was characterized by a string of extremely cold years, with the summers being particularly cool. Many of the marginal northern farms may have been unable to accumulate enough hay for their animals during such summers, leading to inevitable winter hardship and famine. Those who survived may have taken whatever opportunity was available to abandon their farms and move southwards to join relatives in the Eastern Settlement. Archaeological excavation of Western Settlement farms hint at such abandonment, most poignantly in the skeletons of farm-dogs, and in one case a goat, found unburied in the collapsed remains of houses.

The larger and more environmentally favoured Eastern Settlement survived the cold years of the mid-fourteenth century, but the fate of their northern neighbours was a foretaste of what was to come. Moreover, the Greenlanders' most important access to wealth—walrus tusks—had probably been in the hands of men from the Western Settlement. These northerners must have formed most of the parties that hunted walrus in the far north, and these were likely the men who had long-established trading relations with the Tuniit or had learned to trade with the newly arrived Inuit. The loss of this resource must have been a heavy blow to the Greenlanders' attempts to maintain their economic relationships with Europe. This link was already strained by an apparent increase in summer sea ice along the shipping routes to Iceland, as well as by the fact that the Norwegian traders who had traditionally handled Greenlandic produce were being run out of business by the growing sea-power of the Hanseatic League. The German merchants showed no interest in Greenland, perhaps because of the collapse of the ivory trade that may have followed the end of the Western Settlement. Europe's loss of interest in its Greenland colony was best expressed in the fact that the

Greenland Knorr, the official trading-ship sent annually by the Norwegian government, was not replaced after it sank in the late 1360s. A decade later, when Bishop Alf died at Gardar, the church sent no further bishops to Greenland.

Other factors must also have increased the difficulty of life for the remaining Greenland colonists over these years. Ice-core studies show that the Greenlandic climate continued to cool, with series of cold years arriving during the early 1380s and again between 1400 and 1420. Deforestation and the destruction of pastures through overgrazing are known to have had a severe effect on the Icelandic economy, and there is no reason to suspect that the same was not occurring in Greenland. Increasing numbers of Inuit must have been arriving along the outer coasts, and in earlier years would probably have been welcomed as trading partners who had access to ivory and other valuables. But Greenland's isolation from Europe may now have made such trade obsolete, forcing the Inuit to more aggressive means of obtaining the metal objects, wooden boats and other items of European technology that would have become so important to their way of life. The Inuit presence may have done no more than deprive the Norse of seal-hunting and fishing grounds on the outer coast, resources on which they were increasingly dependent as the farming economy suffered. However, the fear of a new, mysterious and potentially dangerous population infiltrating the Norse home-lands may have been an effective motive for individual families to decide to abandon the area.

Most explanations for the disappearance of Norse Greenland have suggested large-scale disasters: Inuit attack, massive crop failure and starvation, epidemic disease or the cumulative effects of generations of inbreeding and malnutrition. No archaeological evidence turned up to date has given any support for such drastic scenarios. The skeletons recovered from Greenlandic churchyards show little indication of disease or malnutrition, and are in fact much like other medieval European populations. There are none of the mass graves or mounds of unburied bodies that would be expected of epidemic victims or the subjects of warfare. In fact, Greenlandic graveyards show that right up to the last days of the settlement, people were buried in an orderly manner, and that they maintained sufficient contact with Europe to follow not only European clothing fashions, but the changing fashions of how the hands of a corpse were placed for burial. Niels Lynnerup, a physical anthropologist whose work has assembled the most complete information on Greenlandic burials, suggests that the end of the colony was much more prosaic: in an atmosphere of slow but inevitable environmental deterioration and increasing fear of attack, individual families simply made the decision to leave their homeland for the greater security of Iceland, Norway or perhaps other European countries. Opportunities to leave may have presented themselves in the form of occasional ships storm-driven west from Iceland, or perhaps with the English fishing vessels that harried the coasts of Iceland through much of the fifteenth century. In the words of the historian Gwyn Jones, it seems likely that "resolute and high-handed English skippers in the fifteenth century sailed into Greenlandic waters for fish and sea-beasts, for honest trade where it offered, and for plunder where it lay to hand."

Kirsten Seaver, a scholar who has recently compiled a fascinating complex of information relating to the end of the Greenland colonies, demonstrates that two voyages that were reported to have been storm-driven to Greenland during the later fourteenth and early fifteenth centuries may not have been entirely accidental. The people on these ships were related to a group of wealthy Icelanders who were also associated with the English fishing enterprise in Iceland. The potential involvement of the remnant population of Norse Greenland in the cod-fishing industry, which was rapidly growing across the North Atlantic and which culminated in the Newfoundland fishery of the later fifteenth century, is an intriguing possibility.

The last that we hear of Greenlandic society came from one of these voyages, a party of Icelanders who were storm-driven to Greenland in 1406 and left four years later. Among the events that they reported was a wedding, performed with complete Christian ceremony at Hvalsey church, the stone walls of which still stand as the best preserved remains of Greenlandic architecture. The next supposedly first-hand news comes in a traveller's tale, reporting how an Icelander known as Jon Greenlander was storm-driven to Greenland about 1540; sailing up a fiord he saw farms and boat-sheds like those of Iceland, and when they landed he found a corpse clothed in a mix of woollen cloth and sealskin, carrying a knife with an iron blade that had been honed to a sliver.

We may imagine the heartbroken loss and grief, the relief and the hope, with which the last families of Greenlanders boarded ship for another life in Iceland, Norway or perhaps England. They were leaving a land where their ancestors had worked and built a European way of life that had survived as long in Greenland as European-based societies have now lived in the Americas. Some must have stayed behind in their beloved homeland, hoping to the last for a good hay-harvest and the return of prosperous times. Occasional small families must have succumbed to hunger in the winter darkness of unheated houses, or died where they worked on the beach, like the man whose corpse Jon Greenlander reported. The next recorded visitor would be Martin Frobisher some thirty-five years later, and he reported the country occupied by Inuit similar to those whom he knew from Baffin Island. A few fortunate children of the last, starved Norse families may have been adopted by those Inuit, who now hunted caribou among the abandoned farms of the fiords and searched collapsing houses for scraps of iron and bronze. After five centuries of existence, medieval Europe's most remote society, established during the heroic age of Viking exploration, had disappeared.

The history and cultural ties of Norse Greenland, however, still resonate in the modern world. Another five centuries have now passed since the Norsemen disappeared, but despite its geographical proximity to North America, Greenland is still, unexpectedly but indubitably, a European state. The link between medieval and modern Greenland is primarily the Church, which from time to time still remembered to worry about its abandoned parishioners. A very dubious letter from Pope Nicholas V to two Icelandic bishops was purportedly written in 1448, requesting that they send a bishop and priests to Greenland, and concerned that most of the churches had been destroyed and the population taken prisoner after a heathen attack that had occurred thirty years previously. There is no indication that this report was taken seriously, or that the Icelandic bishops took any action. In 1492 Pope Innocent VIII appointed a bishop to Gardar, again expressing concern that the Greenlanders had been without a bishop or even a ship for eighty years. As had happened a half-century before, the new bishop never went to Greenland and nothing came of the Church's concern.

By the end of the following century Greenland and its Inuit occupants were becoming known to Europe through the explorations of Martin Frobisher, John Davis, and a host of unnamed English, Dutch and Basque skippers. The Kingdom of Norway (now joined to that of Denmark) never abandoned its claim to its distant possession. However, renewed interest in the lost colony during the seventeenth century can be attributed not to the prospects of sovereignty, but to a long and dreadfully turgid poem published in 1608 by the Danish historian and Lutheran pastor Claus Lyschander. *The Greenland Chronicle* portrayed the Inuit inhabitants described by explorers as the descendants of the medieval Norse. Lyschander urged his King and countrymen to renew their contact with Greenland, and to

Be unaffected by worldly goods
Even though the supplies are small there
You must think of the Glory which will come
From the salvation of many people and the honour of God

He prays that God will

Admonish each and every authority
To turn the mind to Greenland
The people there live like wild beasts
And have forgotten God and Our Lord.
                    [English translation by historian Finn Gad]

These admonitions were buried in the flurry of international whaling activities that occurred in northern seas during the seventeenth century, and in the wars that shredded Europe for much of the period. Lyschander's pleas were not lost, however, and eventually found the ear of an early eighteenth-century Norwegian pastor named Hans Egede. Egede conceived a passion to bring the reformed Lutheran religion to the presumed Roman Catholic remnants of the Norse colonies, and successfully petitioned the King to send him to Greenland. Arriving in 1721, Egede began searching for lost Norsemen in western Greenland, but found only Inuit. Undaunted, Egede decided to bring Christianity to these people, whoever they were, and established a small mission combined with a trading post. The commercial enterprise served to support the expenses of the station, as well as to protect the Inuit from the worst exploitations of the whalers and traders whose ships now frequented the Greenland coast. Europe's re-colonization of Greenland grew from this small and very tenuous beginning. In 1776 the government established the Royal Greenland Trading Company, which held a monopoly of trade and became the main channel of contact between Greenland and Europe. When Norway was absorbed by Sweden in 1814 Greenland became a colony of Denmark, and in this age of colonialism a benignly paternalistic relationship was established, which continued well into the twentieth century.

As in so much of the Arctic, World War II was an important factor in breaking old established patterns. With Denmark under German occupation, the local Greenlandic administration turned to North America for trade contacts and eventually for defence. Treaties allowed the United States to develop airfields to be used for patrolling North Atlantic waters, as well as for the refuelling of aircraft travelling between Europe and America. Although most of the bases were established in sparsely populated areas, the influx of American technology and the related economic effects were significant for the development of the colony. The defence treaty with the United States was replaced by another in 1953 under the banner of the North Atlantic Treaty Organization (NATO), and work began on the massive Thule airbase and radar facility, which patrolled the Cold War skies from northwestern Greenland. In the same year, Greenland's colonial status evolved to that of an integral part of the Danish kingdom. The devolution of powers culminated in 1979 when home rule was pronounced, under which Greenland became self-governing in most matters of social and economic life. Although still a part of the Kingdom of Denmark, contemporary Greenland is in effect a small European nation on the northern margins of North America, much as it was a thousand years ago. The ultimate product of Eirik the Red's violent and adventurous life is the unique mix of Inuit and European cultures that is Greenland today.

6 INUIT

ON A CALM AUGUST DAY IN 1576 Martin Frobisher and his navigator Christopher Hall climbed to the top of a rocky hill on Baffin Island. They hoped to discover the western outlet of the narrow and tortuous strait, which they had been following for days, between lands that they assumed to be Asia to the north and America to the south. They were only a few kilometres from the closed western end of what is now known as Frobisher Bay, but in a haze of distance and hope they thought that they detected two remote headlands and between them the open sea, which "they judged to be the West Sea, whereby to pas to Cathay and to the East India." Looking in the other direction, however, the explorers became aware of a more immediate distraction. As reported in the narrative of Frobisher's voyage published by George Best, "... being ashore, upon the toppe of a hill, he perceived a number of small things fleeting in the Sea a farre off, whyche hee supposed to be Porposes, or Ceales, or some kinde of strange fishe: but comming nearer, he discovered them to be men, in small boates made of leather." Frobisher's crew were about to make their first contact with the Inuit inhabitants of Baffin Island, and to bring the first known description of these people to the knowledge of Europe.

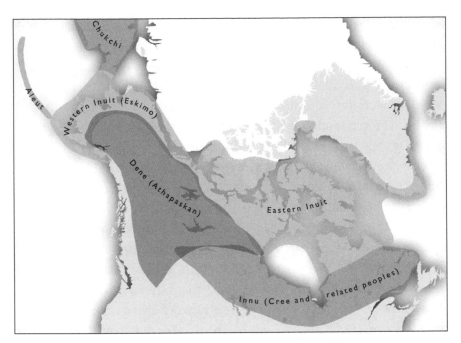

The Inuit and Adjacent Peoples

The English were fascinated by their new acquaintances, who soon boarded Frobisher's ship to trade and to compete with the seamen at gymnastics in the rigging. Christopher Hall reported that "They bee like to Tartars, with long blacke haire, broad faces, and flatte noses, and tawnie in colour, wearing Seale skinnes, and so doe the women, not differing in the fashion, but the women are marked in the face with blewe streekes downe the cheekes, and round about the eyes." Relations between the two groups deteriorated rapidly, with kidnappings and hostages taken in revenge. Inuit captives were taken to England, where they became local celebrities in the brief period before their rapid, and probably inevitable, deaths from disease.

Christopher Hall's comparison of his Inuit acquaintances to Tartars was logical, since he believed that he was on a northern Asiatic shore. That assumption was so widely accepted that the Russian Tsar Ivan the Terrible protested to Queen Elizabeth, demanding the return of Frobisher's captives to their home in his distant Siberian realm. The concept of the Inuit as an Asiatic race persisted even after it had become apparent that Arctic North America lay in a different hemisphere. David Crantz, a Moravian missionary to Greenland and one of the earliest scholars to speculate on Inuit history, wrote in 1820 that the Inuit had originated in "Great Tartary between Mongolia and the Arctic Sea." Crantz conjectured that the Inuit had reached their present homelands through fleeing the expansion of the Mongol hordes, which presumably had pushed northeastwards at the same time that they had invaded medieval Europe.

This view of the Inuit as a people whom one might expect to encounter along Asiatic coasts, and who were recent immigrants to the barren regions of Arctic Canada and Greenland, was a surprisingly accurate first assessment, but was soon to be buried beneath another theory, one that was far more radical and romantic. The explorers and scholars of the nineteenth and twentieth centuries came to believe that the Inuit were the remnants of an ancient Stone Age people who had survived beyond their time in a world cut off from civilization by their intolerable environment. It was this concept of the Inuit past that would stick, and which was reinforced by the romantic and sensationalist exploration literature of the nineteenth century. Most of the world, learned to regard these northern people, who were then known as Eskimos, as a primeval and unchanged race, isolated in their Arctic fastness and shaped through uncounted millennia by ingenious adaptation to the rigours of their environment.

(As a terminological digression, the name *Eskimo* has been said to derive from an Algonkian Indian term meaning "people who eat raw meat." This is an excellent characterization of the people of the Arctic coasts, but the name is currently thought by some to be derogatory and in Canada is usually replaced by the name "Inuit," which simply means "humans" in their own language. Greenlandic Inuit commonly refer to themselves as Kaladlit or Greenlanders as well as Inuit, while the name Eskimo is in general use by the peoples of Alaska and Siberia. Linguists tell us that, in fact, the original Algonkian Indian term probably meant only "the people who live up the coast." "Inuit" is used here to refer to those Eskimos who live in northern Alaska, Arctic Canada and Greenland, all of whom speak a single language known as Inuktituut.)

When European explorers penetrated the ice-filled seas of Arctic North America, they found the land occupied by a people remarkably different from any that they had ever known. They were fascinated by the Inuit's ability to survive comfortably in a land that Europeans found appallingly barren. The essential humanity of the Inuit stood in conspicuous contrast to their way of life, which Europeans saw as totally alien. The English naval explorer John Ross wrote, of his 1830 encounter with the Inuit of the eastern Arctic, that philosophers must interest themselves in speculating "… on a horde so small, and so secluded, occupying so apparently hopeless a country, so barren, so wild, and so repulsive; and yet enjoying the most perfect vigour, the most well-fed health, and all else that here constitutes, not merely wealth, but the opulence of luxury."

There is little doubt that explorers and travel writers accentuated the simplicity, the isolation and the alien quality of Inuit life. The pristine nature of Inuit culture, uncontaminated by contact with Europeans, was an especially appealing aspect of nineteenth- and twentieth-century accounts. Time and again explorers reported encountering groups of Inuit who had never before seen a white man. In 1818, when John Ross met the Inuit of far northwestern Greenland, whom he called "Arctic Highlanders," he claimed that prior to his arrival they had considered their small group of a few families to be the only humans on earth.

Such reports exerted a magnetic effect on the developing science of anthropology. It was an exciting thought: that a secluded Arctic culture would provide opportunities for the study of human social and cultural development in isolation from parallel developments elsewhere. Vilhjalmur Stefansson, an explorer who thought of himself as an anthropologist, described in *My Life with the Eskimo* his 1910 meeting with the Inuit of the Central Arctic:

> *Their existence on the same continent as our populous cities was an anachronism of ten thousand years in intelligence and material development. They gathered their food with the weapons of the men of the Stone Age, they thought their simple, primitive thoughts and lived their*

*insecure and tense lives—lives that were to me the mirrors of the lives
of our far ancestors whose bones and crude handiwork we now and
then discover in river gravels or in prehistoric caves.*

The Inuit soon became the subject of broad anthropological con-
jecture. One hypothesis proposed that the Inuit were an actual
remnant of the ancient peoples who had occupied the Palaeolithic
caves of Ice Age Europe. Others, including the leading scholars on
the subject during the early twentieth century, conjectured that the
Inuit were the descendants of ancient North American Indians who
had been forced northwards by warlike neighbours into regions
where they were compelled to develop a way of life that would
allow them to survive the bleak Arctic environment. The Inuit
became the thrilling archetype of an aboriginal population that was
truly *ab origine*: they were the only people who had ever occupied
this region of the world, they had lived there since time immemorial, and they had developed a uniquely intimate relationship with
the Arctic environment, adapting to it as closely as had walrus and
caribou and polar bears.

This consensus on the nature of the Inuit and their way of life
encouraged travellers and writers to portray them as a very distinctive form of human being, with patterns of thought and reaction
quite different from those of Europeans. In his illuminating book
The Other Side of Eden, which describes the relations between aboriginal hunters and the peoples of more global societies, Hugh
Brody reports a 1973 conversation with his friend and mentor
Anaviapik in the Baffin Island community of Pond Inlet. "How is
it," Anaviapik asks, "that in the old days the Qallunaat [Europeans]
always thought that the Inuit had no thoughts and that we Inuit
were mindless? Is that what you have heard?" Evidence that
Anaviapik's question is based in reality is easily found, and not only
in the journals of the first European explorers to penetrate Arctic
regions. An excellent example can be taken from the work of Jean
Malaurie, a French geographer and anthropologist who passed a year
in northwestern Greenland during the mid-twentieth century. In his
popular 1955 book *The Last Kings of Thule*, Malaurie felt comfort-

able presenting "An approach to Eskimo psychology," which reads as though it refers to an alien and non-human race:

> *It is difficult with exact words to fix these beings with their expressive and mobile faces, and so very unpredictable—without taking the life out of them. Eskimo nature evaporates when one is bold enough to explain it. . . . Like a chameleon the Eskimo leaves a complex memory, a host of varied contradictory impressions. Actually, watching how he lives, you feel he is a marvellous kaleidoscope. Spontaneous and genuine as any-one could be, he wants first of all to satisfy his desires. It is rarely that rational reflection comes to modify, decrease or abort one of his impulses. The only thing that can act as a substitute is a new and irresistible desire. If it is not satisfied the Eskimo loses stature and sometimes becomes really ill. . . . Only concrete life stirs him. Teachers whom I have questioned are unanimous on this, whether they speak of children or adults. The Eskimo mind is incapable of lingering on an abstract idea. Birket-Smith reports that Knud Rasmussen* [both Danish anthropol-ogists of the early twentieth century] *asked a Netsilik one day which was the shorter of two ways. "Although the man knew perfectly well and could say the time it took to go along both, he could not keep them simultaneously in his mind." . . . This people's logic is entirely subjective and emotional. "Instead of being classified in his mind, his ideas follow one another laboriously, one pushing the other or chasing it away" (R. Buliard)* [a missionary in Arctic Canada]. . . . *The Eskimo cannot bear a peaceful life for long. When adventure does not come he creates it.*

And so on. The people whom Malaurie studied were the descendants of the "Arctic Highlanders" whom John Ross had met more than 130 years before. In the intervening period they had been in frequent and in latter years continuous contact with the outside world, and had served as the "Sherpas" who guided and made possible various explorations across the Arctic pack towards the North Pole. While Malaurie lived among these people the immense Thule airbase and radar station was being constructed by the American military in the heart of their country. The extent of their awareness of global

matters is suggested by the response of one of Malaurie's informants
to an ink-blot test: "Two men tearing at a piece of meat... Or rather,
the two men are: the American to the right and the Russian to the
left; they are furious with one another and with all their strength they
are dragging to themselves the world which is at their feet. Under
their effort the world breaks up." There is something incongruous,
even eerie, in the contrast between this consciousness of the mid-
twentieth-century world, and the other-worldly characterization of
"The Eskimo" drawn by Malaurie and his fellow anthropologists
and travellers.

Farley Mowat's books on the Inuit of the central Canadian Arctic,
which were so influential in shaping the public view of the Inuit
during the 1950s, drew from the same intellectual atmosphere and
image of an ancient and isolated Arctic society. To his great credit
however, Mowat did not follow the scholars and missionaries of the
time in claiming that the "Eskimo mind" was different from the
mind of "civilized man"; he saw the effects of ancient isolation only
in the deeply ingrained patterns of life that allowed survival in the
most barren reaches of the Arctic.

The mid-twentieth century also saw the development of the Inuit
art industry, which was to become a major economic force in northern Canadian communities. This venture was promoted as the
sudden blossoming of an ancient and unchanging Stone Age way
of life into artistic creativity when confronted with the myriad,
confusing technologies and societies of the modern world. This
characterization of the art has had sustained success as a commercial concept, and it continues to influence the public's idea of the
Inuit. Few have questioned the historical basis of its claims.

More disquieting is the realization that the concept of a uniquely
adapted people was at the base of government decisions that were to
have grave effects on the lives of Inuit communities. In 1953 the
Canadian government became concerned that American military
interests might not recognize Canada's claim to sovereignty over the
unoccupied islands of the Arctic archipelago. Since international law
interpreted "effective occupation" as a sufficient claim to sovereignty,
the establishment of Canadian Inuit communities on these islands was

The modern village of Qausuittuq in High Arctic Canada was established during the 1950s by resettling people from Baffin Island and the subarctic coasts of Hudson Bay. (Robert McGhee, Canadian Museum of Civilization)

seen as an efficient means of protecting the national interest. The scheme had precedents in other Arctic regions. In 1925 the Danish government had resettled Inuit from the eastern Greenland community of Ammassalik to Scoresby Sund, an unoccupied area 500 kilometres to the north, in order to counter a Norwegian claim to portions of the East Greenland coast. The following year the Soviet government moved a small Eskimo community to Wrangel Island off the northern coast of Chukotka, in order to replace an occupation of Alaskan Eskimos established by American interests. The Canadian effort is the best documented of these cases, as well as the most drastic, since it involved the movement of small groups of families over 1,500 kilometres to the north of their productive Subarctic homeland, into a barren High Arctic environment of intense winter cold and prolonged winter darkness that they had never previously experienced.

In recent years the surviving colonists and their descendants have sought reparation for having been relocated against their will, and the detailed studies resulting from this claim have illuminated

The resettlement to Qausuittuq was encouraged by the existence of archaeological remains of an ancient Inuit village at the location. The whale bone roof supports of this 800-year old house have been recently reconstructed. (Robert McGhee, Canadian Museum of Civilization)

the attitudes of the government officials who planned and carried out the relocations. The justification offered for the move was usually that the Inuit would benefit from it, since they were in danger of becoming overly familiar with, and dependent on, the resources of the European settlements established by traders, missionaries and administrators in their Subarctic home territory, and that by this intervention their ancient way of life could be saved from being contaminated, spoiled and destroyed through contact with outsiders. Rather than accepting the Inuit as citizens of the modern world, the government assumed that they were in need of protection from that world. The prevalent admiration for "unspoiled" Inuit society, and the assumption that this society was the result of an ancient Arctic adaptation, were basic to the belief that Hudson Bay Inuit would have little trouble adapting to the unfamiliar High Arctic environment. The Inuit would rehabilitate themselves from village indigents to noble hunters in the isolation of the High Arctic, while incidentally establishing Canada's effective occupation of this distant region.

The archaeological remains of old Inuit villages lay near the proposed colonization sites, apparently encouraging the officials in their belief that contemporary Inuit would be capable of living by hunting in their new environment. What the officials did not know was that these villages had last been occupied several centuries before, at a time when whale populations had not yet been decimated by commercial hunting, and when a warmer climate and less extensive sea ice meant that whales and other sea mammals were more plentiful. The original occupants of these sites were whalers who possessed weapons and skills that had long been abandoned, or never known, by Inuit groups who lived by other means of hunting. The use of these villages was not part of a long and unchanging Inuit occupation of the Arctic regions, but one brief episode in a much more complex history.

For the past several years I have excavated at one of these old villages, adjacent to the resettled community of Resolute Bay (the name given to the community by the Inuit exiles is Qausuittuq, "the place with no dawn," which gives some indication of their opinion of their new homeland). In working with the inhabitants of this community, and hearing their stories of the bewilderment and hardship experienced by their parents and grandparents, I have learned that romantic notions and erroneous academic theories may have appalling consequences for the people to whom they are applied. As an archaeologist I am dismayed that misunderstood archaeological evidence and flawed theories of the past contributed to such a desperate venture as the relocation of Inuit communities to the High Arctic. A half-century later, archaeological knowledge supports a view of history that brings the Inuit a good deal closer to the rest of the planet's peoples.

Archaeological evidence relating to Inuit history began to accumulate with the Danish "Fifth Thule Expedition" of 1921–24. Led by the Greenlandic scholar Knud Rasmussen, a team of archaeologists and anthropologists crossed the Arctic by dogsled from Greenland to Alaska, excavating archaeological sites and recording the cultural and physical traits of various Inuit populations. Wherever they dug, the Danish archaeologists found the remains of a culture that appeared along the coasts of Greenland and Arctic Canada several centuries ago, and was quite distinct from that of the recent Inuit. They referred to this early way of life as "the Thule culture," since it had first been identified in old houses excavated near the settlement of Thule in northwestern Greenland. The villages that I have been excavating near the community of Resolute Bay were Thule settlements, built and occupied by people of that culture about 800 years ago.

The Thule way of life appears to have been economically richer, more artistically sophisticated, and much more uniform than later Inuit cultures. Thule people were efficient hunters of sea mammals, including bowhead whales which are the largest animals found in northern waters. Their ability to hunt such animals allowed them to settle in permanent winter communities composed of houses built from stone and turf, their roofs raftered with the huge jaw-bones of whales. This maritime-hunting adaptation was based on boats, weapons and tools very similar to those of the Alaskan Eskimos who occupied the coasts of the Bering and Chukchi seas.

Therkel Mathiassen, who was the leading archaeologist of the Rasmussen expedition, saw the Thule-culture people as the ancestors of the Canadian and Greenlandic Inuit. He suggested that they had not developed from ancient times in the northern interior, but had been a maritime hunting people who had recently moved eastward from an original homeland in coastal Alaska, and he dated this movement to approximately 1,000 years ago. But Mathiassen's perceptive interpretation of Inuit history was soon questioned, as archaeologists began to discover the remains of a much older occupation of Arctic North America. When the technique of radiocarbon dating was invented during the 1950s, it became apparent that some of these older settlements were occupied as much as 5,000 years ago. On the sole ground that these early people were hunters who occupied the same environmental region as did the Eskimos, they began to be called Palaeo-Eskimos ("Old Eskimos") and were considered by many to be the long-sought ancient ancestors of the Inuit.

The identification of the Palaeo-Eskimos as Inuit ancestors acquired a political dimension in Arctic Canada during the later

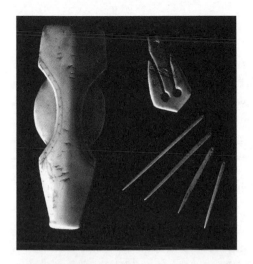

The art of the Thule culture Inuit who moved eastward to Arctic Canada about 800 years ago was usually associated with the decoration of women's tools such as this needlecase and thimble-holder, or objects associated with sea-mammal hunting, such as these snow-goggles for hunting on the spring ice. (Robert McGhee, Canadian Museum of Civilization)

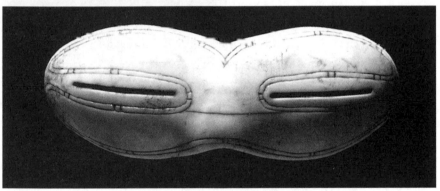

twentieth century, when ancient occupancy was thought to be relevant to the negotiation of land claims. Once again archaeologists found themselves in a situation where academic interpretations were given political significance, and again it has become increasingly clear that the version of history advanced by those with a political interest was likely incorrect. Rather than being ancestral Inuit, it seems probable that the Palaeo-Eskimos can be identified with the people known to Inuit tradition as "Tuniit." The oral history of the Inuit tells that when their ancestors arrived in Arctic Canada they found the country occupied by a strange, rather primitive and non-aggressive people, whom the Inuit soon dispossessed of their lands. As described in Chapter 3, recent evidence from DNA recovered from Palaeo-Eskimo skeletons indicates that these people were not ancestral Inuit, and their identification with the Tuniit of oral history seems to be confirmed.

If the ancient occupants of Arctic North America were not the ancestors of the Inuit, where did the Inuit come from and when did they arrive in the region? Before considering archaeology, we can get a good indication from the study of language. The Inuktituut tongue is spoken across the northern rim of the continent from Greenland to Bering Strait, and the members of the Fifth Thule Expedition who crossed the Arctic in the 1920s found that their Greenlandic speech could be understood as far as northwestern Alaska. There are local and regional dialects, of course, but the range of difference between them has been said to be little greater than that between English as spoken in Glasgow, Bombay and New Orleans. This degree of uniformity can only be the result of a recent dispersion from a single source; it reminds us of the archaeological Thule culture, which also seems to have been the eastward expansion, a few centuries in the past, of a relatively uniform way of life based on Alaskan patterns.

The coastal regions of Alaska and the adjacent eastern tip of Chukotka are occupied by peoples speaking three different Eskimo languages, which are related to Inuktituut but which are not mutually comprehensible. The difference between Inuktituut, West Alaskan, South Alaskan and Siberian Eskimo has been compared to that between English, Danish, Dutch and German. The Eskimo languages, in turn, are related to the language of the Aleuts, the aboriginal occupants of the chain of volcanic islands that trail from the Alaskan Peninsula across the southern edge of the Bering Sea towards Kamchatka. Here the degree of difference has been compared to that between English and Russian, suggesting that the divergence from a common tongue occurred several thousand years ago. The evidence of language would suggest that the most promising place to look for an ancient Eskimo homeland is in the area adjacent to that occupied by their linguistic relatives the Aleuts, and this is indeed where the Eskimo way of life most likely developed.

The southern coast of Alaska is among the most productive marine environments in the world. Bathed by the Alaskan Current sweeping northeastwards from Hawaii, the coast is free of ice and the winters cool and wet rather than frigid. The warm and salty Pacific water mingles with cold, relatively fresh and oxygen-rich Bering Sea water surging southwards through gaps in the Aleutian island chain, creating the conditions that today support the richest fishery in North American seas. Salmon teem in the rivers, marine fish from herring to immense halibut throng the shallow offshore banks, while seals, sea otters, sea lions and whales feed from the bounty. By about 10,000 years ago hunting cultures had begun to exploit the marine resources of the North Pacific coast, and a string of such cultures soon occupied the shores from Japan to California. The ancestors of Eskimos and Aleuts were simply the northernmost of these groups, neighbours of the affluent and artistically celebrated Indian peoples of coastal British Columbia to the east, and of similarly flourishing peoples of Kamchatka and Hokkaido to the west. Much of their maritime hunting technology, as well as many elements of their general culture and way of life, must have been shared along this continuum of peoples occupying the rain-swept shores of the North Pacific Rim.

Archaeological evidence is difficult to obtain in this environment of great distances, heavy vegetation, rapidly changing shorelines and poor preservation of anything other than tools made from stone. The distant ancestors of the Eskimos and Aleuts may have hunted along the southern coast of the Bering Land Bridge, which joined Siberia to Alaska during the low-water episodes at the end of the last ice age. No archaeological remains are known from this early period, and most would have disappeared when sea levels began to rise to flood the Bering Sea about 10,000 years ago. Early sites dating to about 8,000 years ago have been found on Kodiak Island off southern Alaska, and in the eastern Aleutian islands, but their stone tools do not tell us much except that people lived here at the time and were presumably involved in a marine-based economy. By about 5,000 years ago a greater variety of sites are known from the Aleutians and the adjacent southern coast of Alaska, and similarities in the forms of tools suggest that the entire region shared a single culture and economic adaptation. Among the tools recovered are fish hooks, barbed harpoon heads of the type used by recent occupants of the region for hunting sea mammals, and stone lamps for burning sea mammal oil—the latter an essential invention if one wishes to spend winters in the unforested regions beyond the treeline. Although the archaeological sequences are still incomplete, it seems likely that ancestral Aleuts have occupied their island chain continuously since this period, and only slightly less likely that the large Eskimo-speaking population of the South Alaskan coast can also trace a continuous local heritage at least 5,000 years into the past. The central genius of Eskimo culture, however, is based on an undertaking that occurred at some time about midway through this period, when some Eskimo groups adapted the North Pacific maritime hunting way of life to the seasonally frozen waters of the Bering Sea.

The early archaeology of the Bering Sea is patchy, complex and extremely difficult to understand. The period around 5,000 years ago saw the arrival of the people whom we have called Tuniit, Siberians who brought the bow and arrow, tailored skin clothing and the knowledge of Arctic living that allowed them to expand rapidly across Arctic North America from Alaska to Greenland. Early Tuniit sites are found throughout the coastal regions of northern and western Alaska, and there is some indication that they occasionally penetrated the Eskimo and Aleut regions to the south. But while the Tuniit continued to be the primary occupants of Arctic Canada and Greenland until the past few centuries, in Alaska their occupation seems to have been restricted to the tundra regions of the interior after about 3,000 years ago. A diverse and confusing series of finds are known from the coasts of the Bering and Chukchi seas during this period. Some suggest forays to the coast by people of the interior tundras and forests, others Siberian travellers bringing with them the distinctive pottery and other artifacts of northern Asia, still others occasional hunting parties of South Alaskan Eskimos or Aleuts straying far to the north of their homelands.

By about 2,000 years ago some clarity begins to emerge from the confusion. A new level of occupation had now been developed on the coasts and islands around Bering Strait, and this occupation was distinctly Eskimo. Large permanent villages began to be occupied, consisting of houses built from driftwood logs, deeply banked with turf, heated and lighted with lamps burning the oil of sea mammals.

The full paraphernalia of later Eskimo life is represented in the remains of these villages: kayaks and large umiaks, sleds, complicated harpoons and lances, bows and other hunting gear, sewing equipment, ceramic pots and lamps, and an exceptionally sophisticated artistic tradition. This early flowering of the Eskimo way of life has become known as the Old Bering Sea culture.

In the summer of 1995 I had the good fortune to participate in the excavation of one of these ancient villages. The site of Ekven lies on the Bering Sea coast of Chukotka, a few kilometres south of Cape Dezhneva (East Cape). The Diomede Islands, emerging from the drifting pack-ice midway between Asia and America, are clearly visible from the shore, and on a clear day from the hills behind Ekven one can catch sight of Cape Prince of Wales at the western tip of Alaska. Our camp was centred on a weathered wooden hut that had been built and used by the frontier guards who, prior to 1991, had watched this section of forbidden coast where the Soviet Union made its closest approach to America. The guards were now gone, and their hut served as kitchen, dining room and lab for the archaeologists from Moscow's Museum of Oriental Art with whom I was working. The other feature of the camp was a navigation-light for coastal shipping, a fascinating and somewhat frightening remnant of Soviet technology. The light was mounted five metres above the ground on a rickety structure of weathered driftwood logs, and beneath this sagging frame huddled a yellow metal cube containing a small nuclear generator. A geiger-counter showed that the generator was still working, but there was no bulb in the navigation light.

About a kilometre west of the camp, the land behind the gravel beach rises gently over a field of lush grass sprinkled with circular mounds, the remnants of a score of ancient houses. The dwellings are concentrated in a line parallel to the beach, and the advancing coast has chewed its way into the heart of the old village, removing an unknown number of house-remains. For a distance of 500 metres the eroding face of the hillside is a confusion of slumping debris. The whale skulls and driftwood logs from which the ancient houses were built litter the steep slope, mixed with thousands of bones of walrus and seal, potsherds, fragments of carved wood and lost or discarded tools and weapons. The upper edge of the eroding face reveals deposits piled to a depth of three metres, the remains of houses built on top of older demolished houses over a period of several centuries. Nowhere in Arctic Canada had I encountered an ancient Arctic settlement on such a scale. Ekven represents the remains of a community whose population probably numbered in the low hundreds, and the piled debris indicates a continuity of occupation extending through ten centuries or more.

Aside from deciphering the eroding remains of the village, the archaeologists from Moscow concentrated their excavation on the cemetery that lies atop a hill a few hundred metres behind the settlement. Here they have recovered the remains of more than 300 burials, many of them accompanied by a rich assortment of tools, weapons and more esoteric objects. The disparity between meagre and richly furnished burials clearly indicates that the ancient inhabitants of Ekven lived in a society marked by social distinctions: rich and poor, leading families and hangers-on, probably slave-owners and slaves captured in warfare from adjacent groups. Mikhail Bronshtein, the Moscow archaeologist who was my particular mentor at Ekven, has devoted much effort and ingenuity to studying the lavishly decorated ivory weapons and other carvings recovered from these burials. The Old Bering Sea people were skilled at the sculpting and fine engraving of ivory, in a variety of styles, an art that transformed hunting gear and other utensils into a vast range of fantastic animal-like forms. The results of Misha's detailed analysis made obsolete the old assumption that these styles succeeded one another through time. Instead, he proposes that these elaborate styles were the identifying marks of distinct but contemporaneous ethnic communities, some of whose members lived at Ekven—as marriage partners, visiting merchants, perhaps as captives in warfare—or were at least buried at Ekven together with the local inhabitants. This picture suggests a social complexity that transcended the local village level and enmeshed the entire Bering Strait area in an intricate web of ethnic societies, all participating in the same rich economic pattern.

The carved and engraved ivories from Ekven burials provide a clue to the explanation of why this major cultural development occurred in the centuries around 2,000 years ago, and on what it was based throughout the following millennium. These elegant objects were not formed with stone tools, but with cutting and engraving blades made from iron. Indeed, it seems likely that iron-bladed tools were used to carve most of the ivory, antler and wooden artifacts of early Eskimo technology. Although the beginnings of the Old Bering Sea culture are uncertainly dated, they appear to be roughly contemporaneous with the spread of the Han Empire, during which Chinese political and economic control was for the first time extended over an area comparable to that of today's China. The Han period also saw the introduction of iron weapons and tools into central Asia, Korea and Japan during the second century BC. Shortly after this time iron tools began to reach the Eskimo villages at Bering Strait, probably through trade networks that supplied ivory, furs and other Arctic valuables to the burgeoning market in luxuries required by the ruling classes of China and adjacent areas. Whether such a trade was organized through interior trade-fairs, like the one on the upper Kolyma River that flourished during the eighteenth century, or was carried out through a network of coastal peoples whose products eventually reached itinerant Korean and Japanese traders, it seems likely that early Eskimo culture developed in a distant relationship with the economic centres of eastern Asia. A marble spindle whorl found in an Ekven grave, of the type used at the time in the Primorye region on the boundary between Russia and Korea, suggests that the trade followed a coastal route.

If the trade in metal was an important element in the Old Bering Sea economy, it would explain why the largest and richest populations of the time were clustered in a relatively small area: the eastern coast of Chukotka close to Cape Dezhneva, the islands in Bering Strait and Alaska's Seward Peninsula adjacent to Cape Prince of Wales. The communities that occupied these strategic locations would have been in a position to control the entire trade in iron between Asia and North America. We do not know the extent of this trade, as iron preserves very poorly in most archaeological deposits, but a recent chance find provides a hint. A few years ago, the frozen body of an aboriginal hunter was found melting from a glacier in the high mountains of northern British Columbia; the man seems to have died about a century before Christopher Columbus reached the New World, and therefore before European iron-based technology was introduced to most of North America. Among the possessions of the man in the glacier was a small iron-bladed knife, not very impressive evidence in itself, but if one considers the statistical likelihood that a random hunter from a relatively poor interior Subarctic tribe carried an iron knife, the find suggests that the material must have been widely used throughout northwestern North America at the time.

Given the rapid and extensive spread of iron when it was later introduced from European sources, and given the obvious wealth of the early communities about Bering Strait, we may suspect that significant quantities of the precious material were reaching the New World as much as 2,000 years ago. The Old Bering Sea people, the earliest group that we can certainly identify as Eskimo, seem to have built not only a surprisingly rich hunting society exploiting the abundant resources of the Bering Sea, but a complex society of entrepreneurs engaged in what must have been an extremely profitable and competitive commercial enterprise.

The village of Ekven, like most of the other large communities around Bering Strait, seems to have been occupied continuously for a period of between 1,000 and 1,500 years. The Old Bering Sea culture flourished during the first half of this period, and the transformation to a later cultural form was marked by the introduction—at some time between AD 500 and 1000—of two important elements. One was the development of effective techniques for hunting bowhead whales. Bowheads are the largest animals in Arctic seas, relatively slow-swimming and placid compared to the small but dangerous grey whales hunted by Old Bering Sea people. A single bowhead whale could provide fifty tons or more of meat and blubber, and the ability to capture and butcher such an animal encouraged the development of whaling villages at points of land where the annual bowhead migrations passed close to the coast. The largest villages, with several hundred occupants, were established along the

northern coast of Alaska where northward migrating whales could be
hunted in narrow shore-leads in the sea ice of early summer. Bowhead
whaling probably also encouraged Eskimo expansion into the western Canadian Arctic where the whales spent the summer before
returning southwards to the Bering Sea.

The second important development during this period was the
appearance of the recurved bow, based on the Mongol pattern that
at this time was being carried westward to Europe by invading armies
from central Asia. Strengthened by a thick cable of twisted sinew that
supplied most of the bow's force, it was a much more powerful
weapon than earlier bows used in the area. It seems to have been used
against humans as well as other animals, since slat body-armour based
on Asiatic patterns of metal armour, but made from ivory or bone,
begins to appear in the archaeological deposits at the same time as
the bow. Together, bow and armour indicate that the patterns of
inter-community hostility that were observed later, when Russians
began to penetrate the area, have deep roots in the past. The early
Eskimo communities around Bering Strait were obviously not only
willing and able to protect themselves and their valuable commercial
activities, but were capable of expanding into areas occupied by less
well-armed and less aggressive peoples.

It was against this background that the next major development in Eskimo history occurred: the expansion of the ancestral
Inuit eastwards from Alaska to most regions of Arctic Canada and
Greenland. This event used to be thought of as the "Thule migration,"
and seen as the spread of whale-hunters into a new environmental
zone that had become available to them when the climate warmed
throughout the northern hemisphere around 1,000 years ago. My
doctoral dissertation, which I wrote and defended over thirty years
ago, was based on that thesis, which I now simply no longer believe.
The concept of a Thule migration presumed that the Inuit were an
adaptive species, and that like the caribou or walrus they would auto-
matically take advantage of a new environmental opportunity. But
when a little thought is given to this subject, it becomes apparent that
human beings simply do not behave in such a deterministic manner.
Moreover, we now know that the warming climates of 1,000 years

This painting by the nineteenth-century Greenlandic artist Aron of Kangeq depicts Inuit oral historical tradition regarding an early meeting between Inuit and the mediaeval Norse farmers of Greenland. (From Eigil Knuth, *K'avdlunâtsianik, The Norsemen and the Skraelings.* Det Grønlandske Forlag, Godthab, 1965)

ago didn't affect all Arctic regions, and that the climate had begun to deteriorate again well before the Thule people appeared in the regions east of Alaska. Even at their most benign, the sea-ice conditions of the Central Arctic still formed a 1,000-kilometre barrier to bowhead whales and other animals dependent on open water, and created a wide and inhospitable gap between the rich hunting resources of the Western Arctic and the similarly endowed eastern regions adjacent to Baffin Bay. Whatever process brought ancestral Inuit to the Eastern Arctic, it is unlikely to have been the slow expansion of hunting groups across a relatively uniform environmental zone.

I now suspect that the origin of the Inuit was tied to a much simpler motive: the discovery that iron and other metal was available in the Eastern Arctic. The Tuniit, the aboriginal people who occupied most of Arctic Canada and Greenland, almost certainly came into contact with the Eskimo whalers who moved into the coasts of the Beaufort Sea around 1,000 years ago. By this time the Tuniit were carving their celebrated small sculptures, using iron tools made with metal obtained from several meteorites that they had recently discovered in the Cape York region of northwestern Greenland. The Tuniit also exploited the native copper deposits in the Coppermine River region of the Central Arctic, and both meteoric iron and native copper were widely traded. Archaeologist Pat Sutherland has recently demonstrated that the Tuniit of Baffin Island and Labrador were in contact with the Greenlandic Norse, and although little smelted metal has been found in Tuniit sites, these contacts provided another potential source of valuable iron, copper and bronze. Such sources, in the hands of a relatively small, scattered and poorly armed population, may have been attractive enough to motivate Inuit adventurers on journeys of exploration across the barren channels of the Central Arctic. These voyages would have provided the information that the Eastern Arctic had animal resources similar to those of Alaska; that the Cape York meteorites were a source of iron that was available for the taking; and also that the area was visited by *qadlunaat*, blue-eyed strangers from whom smelted metal could be obtained either by trading or through attack. This knowledge may have been the trigger that launched the Inuit migration to the Eastern Arctic.

This postulated motive for Inuit expansion is supported by the fact that the earliest known Eastern Arctic Thule occupations are found along the extreme High Arctic coasts adjacent to the Cape York area, and that these occupations are associated not only with meteoric iron but with quantities of metal and other materials obtained from the Greenlandic Norse. Intriguingly, the styles of the artifacts used by these early Inuit venturers are identical to those used by peoples living around Bering Strait, instead of those used in the whaling villages of North Alaska which would seem to be the most likely base for eastward exploration. The evidence found by archaeologists Karen McCullough and Peter Schledermann, whose excavations at these High Arctic sites have recovered the most comprehensive collections of early material, rather convincingly hints at a rapid movement of people directly from Bering Strait to the shores of Baffin Bay. The radiocarbon dates associated with these extremely early Inuit sites, as well as those from the old villages near Resolute

Bay at the threshold of the Eastern Arctic, suggest that this eastward movement occurred at some time during the twelfth or thirteenth centuries AD. This was the period when the Greenlandic Norse colonies were at their most prosperous, when the Norse demand for ivory and furs to trade with Europe was at its height, and when the Norse would have been most interested in establishing trading relationships with the newcomers.

Having penetrated the Eastern Arctic and obtained control over the metal sources that were a necessary basis for their technology, the Inuit gradually spread throughout the region. Their oral historical traditions state that they either killed or drove away the Tuniit who had been the aboriginal occupants of the area, and these accounts accord with the archaeological fact that signs of Tuniit occupation disappear after the thirteenth or fourteenth century AD. The Inuit established permanent winter villages throughout the region, similar to those occupied by their ancestors in Alaska, and developed a way of life based on the hunting of bowhead whales where they were available. The remains of these Thule culture villages, their winter houses heavily raftered with the jawbones of whales, are scattered along most coastal regions of Arctic Canada and in northwestern Greenland. We still don't know when the Inuit penetrated the areas of southern Greenland that were occupied by the Norse, but both archaeology and oral traditions appear to indicate that the two populations co-existed in the area for some time, and that the Inuit occupation continued after the disappearance of the Norse. The people encountered by Martin Frobisher in Greenland and Baffin Island were Thule culture Inuit, and their way of life—based on the hunting of bowhead whales and other sea mammals, and the occupation of permanent winter villages of heavily insulated houses—was not dissimilar to that which their recent ancestors had brought eastward from Alaska.

A much different picture of the Inuit way of life was painted by the Europeans who began to penetrate the channels of the Canadian Arctic during the eighteenth and nineteenth centuries. By this time the High Arctic coasts along Lancaster Sound—the route by which the Thule people first moved eastward from Alaska—had been abandoned. The only people left in the High Arctic were the tiny

When Europeans began to penetrate Arctic Canada they found the villages of the Thule whalers abandoned. Through much of the central Arctic, small groups of Inuit lived a much more nomadic way of life than had their ancestors a few centuries earlier. (Robert McGhee, Canadian Museum of Civilization)

band who called themselves Inughuit, Ross's "Arctic Highlanders," who made a tenuous living centred on the bird-nesting cliffs of northwestern Greenland. Elsewhere the only signs of humanity were the deserted remains of winter villages, the whale bone roofs of winter houses collapsing under winter snow, and lush moss rapidly covering middens dense with the bones of animals that had once supported a thriving population. Similar deserted villages lay along the coasts of the more southerly Arctic islands and of Hudson Bay, but descendants of the people who had once occupied these communities continued to live in the region. These were the people whose way of life was to form the stereotype of Inuit culture: tiny scattered bands of families hunting seals and caribou, living close to the edge of starvation in tents and snow-house winter villages built on the sea ice.

What had happened to the whale-hunting Alaskans who had moved eastward across Arctic North America a few centuries before?

How had they dwindled into these small and economically marginal societies? The first instinct of historians and archaeologists has been to turn once again to an environmental explanation: the onset of colder climates and an increase of sea ice, which made the Thule way of life impossible in most regions of Arctic Canada. The weather deteriorated across much of Europe between the sixteenth and mid-nineteenth centuries, a period that has become known as the Little Ice Age. But it has now become apparent that the environmental effects of climatic change varied greatly from one region to another, both in their extent and in the time period when they occurred. In some parts of Arctic Canada episodes of climatic cooling occurred as early as the twelfth or thirteenth century, when Thule culture was at its height, while other areas produced no evidence of environmental change until centuries later. Another approach to explanation has suggested that the massive Basque whaling effort in Labrador waters during the sixteenth century depleted the whale stocks on which the Thule Inuit depended. But when European whalers penetrated Arctic Canada during the early nineteenth century, at the height of the Little Ice Age, they found large populations of bowhead whales still migrating past abandoned Thule villages, so neither extreme ice conditions nor commercial whaling seem to have been a cause for the disappearance of the early Inuit whale-hunting economy.

Once again, our attempts to explain a major change in the Inuit way of life may have been biased by our tendency to think of the Inuit as an isolated people whose life was ruled by environmental conditions, and to ignore the influence of contacts and engagements with other nations. The sixteenth-century cessation of Inuit occupation in the High Arctic, for example, approximately coincides with the disappearance of the Norse Greenland colonies, which may have served them as a continued source of trade. At the same time the appearance of European explorers, whalers and traders along the coasts of Baffin Island and especially of Labrador, and later around the coasts of Hudson Bay, may have been a magnet drawing the metal-hungry Inuit southwards. The Inuit whom Martin Frobisher met in Greenland and Baffin Island in 1576–78 were apparently already familiar with Europeans, and were in possession of European

This Inuit snowhouse village was photographed in 1915 in the central Canadian Arctic. Although such villages are often considered typically Inuit, they were in fact only used among the small groups occupying the central Arctic. Most of the larger Inuit populations to the east and west built substantial houses of logs, stones and turf to provide winter shelter. (Canadian Arctic Expedition, Canadian Museum of Civilization Archives)

goods. We may suspect that unrecorded voyages to these regions occurred at least occasionally, perhaps as an offshoot of the massive sixteenth-century European whaling and fishing efforts in southern Labrador waters.

The continuing importance of metal and other European materials in Inuit technology may have had a significant effect on the distribution of populations, and on the annual round of economic activities that they followed in order to get access to this resource. Sixteenth- to eighteenth-century European contact along the coasts of Greenland, Labrador, Baffin Island and even Hudson Bay led to an increasing involvement of Inuit in trading activities, hunting or trapping for trade, and eventually direct employment in the trapping and whaling industries of the nineteenth century. Disease may have been another factor in breaking down the Thule Inuit way of life. As happened among the Indian populations of eastern North America, the agents of disease must have accompanied the metal trade into regions far distant from those where direct contact with Europeans had occurred. The small populations and dispersed settlements of the Inuit probably prevented the occurrence of epidemics as widespread as the smallpox that devastated the Dene Indians to the west of Hudson Bay during the late eighteenth century, or the early-twentieth-century measles epidemic that almost destroyed the Mackenzie Delta Inuvialuit, but a continuous series of small-scale outbreaks must have plagued Inuit society over the past five centuries.

As in the case of the disappearance of Norse Greenlandic society, the disintegration of the Thule Inuit culture must have resulted from several interacting causes. The unpredictable effects of climatic change may have been compounded by choices made by local groups: for example, a decision to change their patterns of seasonal movement or hunting activities in order to acquire European goods. The neglect of summer whaling meant the inevitable abandonment of permanent winter villages, which were dependent on food stored from the summer hunt. The hardships of travelling and living in temporary camps must have been intensified by waves of illness brought by returning traders, and eventually by malnutrition when foreign foods began to be used by people whose hunters spent much of their time trapping fox or hunting whales for profit.

The image that the world holds of a people, and indeed a people's own self-image, is closely tied to the understanding of their history. The image of the Inuit as a simple, unspoiled, unthinking society of noble and capable hunters is intimately related to the view of Inuit history as an ancient and isolated adaptation to a bitter environment. It is now clear that the Inuit were not the inheritors of such a simple and unlikely cultural tradition. Rather, the way of life described by nineteenth-century explorers had developed during the preceding few centuries in response to rapidly changing economic circumstances. Far from being the most isolated population on earth, I believe that the ancestors of the Inuit will prove to have been an entrepreneurial North Pacific people who were attracted to the Eastern Arctic during the thirteenth century by the prospect of trade with Europeans, and whose way of life has

developed in contact with the evolving global culture and economy ever since. The perspective on Inuit culture that was
presented by John Ross or Jean Malaurie, which allowed the
Canadian Government to undertake the relocations of the 1950s,
and is still promoted by the Inuit art industry, is truly obsolete.

The realization that the Inuit are not a peripheral people was
forced on my mind one night on the coast of Chukotka, as I climbed
by myself over the remains of the ancient community at Ekven. A
few kilometres up the coast, the low night-time sun was throwing an
orange glow on the rocks of Cape Dezhneva, the most easterly point
of Asia, and on Great Diomede Island halfway across the Bering Strait
to Alaska. In the bright calm night I suddenly had the overwhelming
sense that I was not standing at the distant margin of a world, the end
of the earth, as far as one could travel from Europe. Instead I was
standing at the very heart of another world, a nexus that for millennia has linked the peoples and cultures of Asia and America. It was a
world in which many nations and cultures had flourished, among
them the Inuit and their way of life.

7 ICE AND DEATH ON THE NORTHEAST PASSAGE

THE PAST FIVE CENTURIES OF ARCTIC HISTORY are usually presented as a chronicle of exploration. History becomes a litany of the deeds of those (usually) men who travelled beyond the bounds of their own known world, returning with tales of hardship, adventure and amazing discoveries. Arctic geography becomes a gradually expanding "known world," the advancing frontier of Europeans' increasing awareness of the general outlines of coasts and rivers, routes of travel through ice or storm, and the animals and primitive tribes that occupied the country. Maps from these years of exploration depict the regions beyond this frontier as either snowy blanks or territories provisionally sketched with icy coastlines, towering mountain ranges, open oceans and strange peoples.

The term "Arctic explorer" conjures the dramatic image of a heavily bearded, frostbitten figure pushing through an endless blizzard, surviving on pemmican, willpower and the quest for knowledge or for fame. This specific image comes from Victorian travel literature, a genre that was invented or perfected by travellers such as Charles Francis Hall, an explorer in the classic mode who presented himself as the hero of some outstandingly well-narrated adventures. The centuries before and since the nineteenth have seen a great vari-

The Northeast Passage

ety of Arctic explorers, many of whom bore no resemblance to the classic image, who found themselves in Arctic regions on a diversity of missions and with many different motives. The royal favour that would attend the discovery of a passage to Asian markets, the immediate fortunes to be made from discovering gold or whales, the advancement of a career through dutiful execution of orders, the public fame to be gained from distant achievement—all drew the men who would be known as explorers.

The passage of time has veiled the competitive striving and personal ambition that drove most of these projects, giving them instead a romantic aura of high-minded adventure. In a similar way, time has lent an appearance of solidity to the purpose, the organization and the actual execution of these enterprises. Most exploration accounts paint a picture of participants committing themselves to well-planned projects for reasons that were thoroughly deliberated; of companies of men selected because they possessed the skills and capacities that would be essential in the circumstances expected; of equipment and supplies rationally acquired and suited to the task. All of these conditions may have applied to some Arctic expeditions. But the number of ventures that encountered unnecessary hardships and ended in tragedy or farce suggests that poor judgement and slipshod planning may have been the norm rather than the exception. Layers of vague reporting and optimistic historical assumption have disguised the deficiencies of these episodes. Even Charles Francis Hall, the American journalist turned Victorian Arctic explorer, went to his Arctic grave not from the rigours of exploration in a cruel environment, but from arsenic administered as the cure for a personality conflict between the famous explorer and his long-suffering physician.

The tireless heroism and meticulous, unquestioning commitment to duty so often ascribed to Arctic explorers are unimaginable and often frankly incredible. I find it easier to understand the fatal indecisiveness that cost Henry Hudson his life at the hands of mutineers; the sheer lack of planning that caused the inevitable disaster suffered by Hugh Willoughby's crew on the bleak coast of the Barents Sea; and the class-bound arrogance that made Sir John Franklin someone best avoided when he asked for your company on a trip to distant and lonely places. Arctic exploration has produced unquestionable examples of heroism, survival against all odds, and the endurance of terrible hardship in pursuit of duty or an ideal. But the story of Arctic exploration, which has so often been cast as a simple tale of individual achievement, is far more than that. It is a narrative spun from the terror of being locked deep in the heart of a lethal and alien environment; the dreadful tedium as months of inactivity drag by in cramped and uncomfortable quarters; the homesickness that would barter life for the sight of a country garden; and the bleak depression that settles on those whose lives have been reduced to an apparently endless sentence of hard labour in a world of wind and ice. The achievements of Arctic explorers are best appreciated in the realization that they were not accomplished in an atmosphere of superhuman competence, but rather in the muddle of frailties and ambitions and failure that characterize most other aspects of human activity.

The age of Arctic exploration began with the Renaissance in learning and commerce that originated in Italy during the fifteenth century, but which took a considerable time to reach the shores of northern Europe. By the time that the English and the Dutch began to develop commercial interests in overseas trade and royal ambitions for foreign dominion, most of world exploration and trade had fallen firmly into the hands of southern European powers. The early sixteenth century saw the establishment of Spanish colonies and enterprises in the New World, and the discovery of Magellan's strait which led to the gold of the Moluccas and the wealth of the Indies. Portuguese ships had developed the African trade in gold, ivory and slaves, and had rounded Cape Horn to reach the vast markets of India. Even the region of North America known as the New Found Land, to which the English had sent the Venetian pilot John Cabot in 1497, was exploited almost entirely by Basque whalers and French cod-fishers.

Lacking both the maritime proficiency and the political will to challenge the southerners, but fired by the fortunes that could be made in distant parts of the world, the Dutch and the English began looking for alternative routes to these riches. Knowledge of geography and of the techniques of navigation became valuable commodities in the north, and men who possessed such knowledge soon found themselves in positions of importance and potential wealth. Two such men arrived in England around 1550, and between them laid the groundwork for the explorations that led to the European discovery of the Arctic regions.

The first of these men was Sebastian Cabot, son of the Venetian who had led the 1497 English expedition to the New Found Land. As a very young man, Sebastian may have accompanied his father on this voyage, and at times took for himself the credit for its discoveries. His rather shadowy history may also have included another voyage in 1498, during which his father appears to have been lost. This could have been the basis of his later claim to have made an

Arctic voyage in the same region, and to have reached a high latitude in ice-free seas before being forced to turn back by a frightened crew. He had spent most of the intervening decades in the service of Spain, and when he arrived in England in 1548 he brought a wealth of geographical knowledge. Cabot's reputation has suffered among historians, many of whom consider him to have been nothing but a jovial fraud, but the history of early sixteenth-century exploration is so poorly documented that the man may actually have undertaken some of the ventures that he claimed. In England his reputation was sufficient to convince shrewd merchants and practical sailors that an ice-free Arctic Ocean existed, and that this ocean could lead them to the wealth of oriental lands.

The second man to arrive in England at this time was John Dee, an Englishman who had studied on the Continent with the mathematicians and cosmographers of Paris and the geographers of the Low Countries. Dee came to be thought of as the most knowledgeable man in sixteenth-century Europe, a polymath whose interests extended from the geography and geopolitics of the secular world to the realms of the spirits and communications with angels. He served as astrologer to the young Princess Elizabeth, and later served her as Queen by developing and promoting the concept of a British Empire based on both mythological and contemporary English explorations of the northern parts of the world. In the 1550s Dee brought to England the geographical knowledge of the polar regions that he had learned from the great Flemish geographers Gerhard Mercator and Abraham Ortelius, both of whom postulated the existence of a navigable Arctic Ocean. He also brought navigational instruments from the Continent, and assiduously instructed (often with little result) a generation of English sailors in their use.

The first English ventures were directed northeastwards in the search for a navigable passage to the north of Asia. Given the assumptions of the time regarding the geographical arrangement of the world, it must have seemed quite likely that such a route would be found. The tales of sailors such as Sebastian Cabot supported ancient beliefs in the existence of an open polar ocean. Moreover, arguments of divine symmetry in the creation of the world implied that navigable channels should exist to the north, as well as to the south of the world's continents. When Martin Frobisher later mistook the narrow inlet of Baffin Island's Frobisher Bay for the Northwest Passage to Asia, he was clearly thinking of a northern equivalent to the strait that Ferdinand Magellan had discovered fifty years before at the southern tip of the Americas, a tortuous and dangerous channel 300 kilometres in length separating the mainland from Tierra del Fuego.

Similarly, North Cape on the coast of Norway could be considered a northern equivalent of Africa's Cape of Good Hope, and after rounding that promontory one might well find a broad and open passage to the coasts of Tartary and Cathay. The cape at Europe's northern extremity is in about the same longitude as that at Africa's southern tip, and eastward from both capes the coasts trend back towards the temperate zone. However, instead of finding a northern equivalent of the Indian Ocean, the traveller rounding North Cape encounters the gray Barents Sea, where the rapidly cooling waters of the North Atlantic Drift contend with the ragged edge of the polar pack-ice that constantly impinges from the north and east. The Barents Sea is protected from the worst effects of the ice by the 1,000-kilometre spine of Novaya Zemlya, a northward extension of the Ural Mountains barely separated from the mainland coast. To the east of this huge island lies the Kara Sea, more than 1,000 kilometres wide and protected from the polar pack only by offshore winds and the warm outflow of two of Siberia's massive northward-flowing rivers: the Ob to the west, and the Yenisey to the east. These forces create open water conditions during the late summer through most of the Kara Sea, but to the east of the Yenisey the pack-ice locks directly against the coast of the northward-jutting Taimyr Peninsula. Beyond this bleak promontory lies a further 4,000 kilometres of ice-jammed waters—the Laptev Sea, the East Siberian Sea and the Chukchi Sea—before a ship can find her way southward through Bering Strait and into the temperate waters of the Pacific Ocean. The reality of the Northeast Passage was far more difficult than envisioned by the theoretical geographers of the sixteenth century.

I have not experienced the Northeast Passage, but have seen both the eastern and western approaches and can attest that they do not promise ease or comfort. Both are cold and foggy seas during the brief navigation season of late summer, with grey wind-whipped waves crawling under a low grey sky. From the deck of a small ship off the east coast of Svalbard I watched the eastern sky turn white, the reflection of nearby pack-ice on a low blanket of cloud over the Barents Sea. Years later I watched the same ice-blink over the Chukchi Sea from the northern coast of Chukotka, a few kilometres to the west of Bering Strait. For much of the intervening 5,000 kilometres the sea is a realm of ice, a substance both mysterious and feared by early European navigators who encountered it for the first time.

The ice of the polar basin exists in a variety of forms. It begins to develop in September and October of each year, when temperatures drop well below zero and the surface of the sea smokes as the last heat is drawn from its surface to dissipate in the frigid air. The sea develops a hazy greasy appearance as tiny crystals form and spread, eventually coalescing into a thin elastic layer that undulates in the swell but, only a few centimetres thick, can support the weight of a human or a seal. My only experience of such ice was enough to tell me that I needed much more knowledge before I could feel safe on its surface. This ice gradually thickens, breaks and reforms in jagged patches, incorporates snow or rain fallen on its surface, and eventually develops into white pack-ice from one to three metres thick. The pack-ice that forms in coastal regions is locked to the shore, relatively stable and unmoving, and is known as land-fast ice. By the following spring most of the salt has leached from this ice, and its brilliantly white surface is mottled with azure ponds of melted water. Stable, level and smooth, this ice is a delight on which to travel, hunt or play. As spring advances it becomes veined with narrow leads of open water, home to seals and other mammals, and in June or July disintegrates into chunks of blue-white pack-ice drifting with the wind in widening stretches of open water.

Beyond the realm of land-fast ice is the free-floating polar pack, a mix of new ice and refrozen multi-year ice three to five metres thick, constantly in motion as surface winds and subsurface currents propel it in a slow endless loop around the Arctic basin. The polar pack occasionally incorporates icebergs calved from the glaciers of

Midsummer sea ice conditions on the polar pack. (Robert McGhee, Canadian Museum of Civilization)

Arctic islands, or ice-islands hundreds of metres thick, broken from ice-shelves formed along Arctic coasts during the last ice age. The continuous movement of the pack opens long leads of water, lakes and temporary seas form at random and suddenly close as the adjacent shores crush and over-ride one another to form massive pressure-ridges. The polar pack renews itself each year, as the edges flow southwards through outlets into warmer waters or grind against obstructing coasts. From a cliff on a far northern island I once watched this permanent cap of floating ice as it encountered a rocky shore. Driven by the power of winds and currents sweeping across the pole, ice as hard as concrete and as high as a house is broken, tilted, piled, ground and refrozen into an impassable chaos, booming and groaning as if it were alive and suffering. The wooden ships of the age of exploration had no chance against such a force, and could only contemplate a passage by gaining sufficient knowledge to avoid the ice. This knowledge came through expending hundreds of years of effort, and at least as many human lives.

The first European known to have rounded North Cape was the

Norse chieftain Ottar, who boasted of his adventure while he visited
King Alfred of England around AD 890. Ottar described himself as
the most northerly of Norsemen and probably lived well above the
Arctic Circle, where he traded with or exacted tribute from the
Saami hunters and fishers who lived further to the north. Ottar
claimed to have sailed northwards until the Norwegian coast turned
east, then sailed nine days to the east and south, following the coast
into what must have been the White Sea. Here he encountered many
walrus, which he killed for their tusks and hides, and a great river
(perhaps the Dvina, at the mouth of which Arkhangelsk was later
built) inhabited by settled people called Permians. Similar expeditions for hunting or for trade with the peoples of northern Russia
appear to have continued during subsequent centuries. A merchant
adventurer named Uleb from Novgorod claimed to have penetrated,
in AD 1032, the strait to the south of Nova Zemlya, and explored
the coast of the Kara Sea.

Over the following centuries the coasts of the White Sea saw the
development of the Pomor people, Russian sea-farers, maritime
hunters and fishers who developed the early trade in furs and other
products of the hunt. The ships and navigational skills of the Pomors
probably benefited from contacts with Norse sailors like Ottar, but
seem to have been developed mainly through their commercial
activities across the system of huge lakes and rivers of Karelia linking
the Baltic to the White Sea. The knowledge and Arctic maritime
skills gained by the Pomors were invaluable to the explorers who
later ventured in search of a passage to the north of Eurasia, and their
trading contacts sufficiently important that King Hakon of Norway
built a fort and the Archbishop of Trondheim consecrated a church
on Vardø Island in 1307. Vardø is located about 200 kilometres to the
east of North Cape, where the coast turns south into Varanger Fiord,
and seems to have marked the boundary between Norwegian and
Russian lands. The settlement that grew up around the fort
(Vardøhus, which became Wardhouse to early English voyagers) long
remained a destination for northern venturers. In later centuries
Wardhouse was the last port of call for Russian Pomor hunters travelling to Spitsbergen from the White Sea coast; here they traditionally

drank up their advances of pay, departing ill and in debt for a dreadful winter's work in the darkness of the High Arctic.

By the mid-sixteenth century, with Spain and Portugal reaping astonishing benefits from their ventures in distant parts of the world, English merchants realized that if they decided to face the considerable risks involved, they too could find markets for their cloth and other goods, and gain windfall profits from the importation of foreign merchandise. Lacking the royal backing afforded the merchants of other countries, they formed a new type of organization: the joint stock company. The first such association was the "Mysterie and Companie of the Merchants Adventurers for the Discoverie of Regions, Dominions, Islands and Places Unknowen," which later came to be known as the Muscovy Company. Sebastian Cabot was named Governor of the Company, and more than 200 London merchants and noble speculators pooled their capital to finance a fleet of three vessels that set out in 1553 to round North Cape and explore the route to China across the northern coast of Eurasia.

The ships—*Bona Esperanza, Bona Confidentia* and *Edward Bonaventure*—were confidently named, and their hulls were sheathed in lead to protect them from teredo worms in the tropical waters that they planned to reach. The commander of the expedition was Sir Hugh Willoughby, a man of good birth and military reputation but no nautical experience. This was to be compensated for by the expertise of the fleet's pilot, Richard Chancellor, and the master of the *Edward Bonaventure*, Stephen Borough. However the *Edward Bonaventure*, with both Borough and Chancellor aboard, was separated from the others during a storm off northern Norway, and this unfortunate accident probably sealed the fate of the other vessels. Willoughby, the least able of the expedition's principals, led the remaining two ships boldly northeastward across the ice-free waters of the Barents Sea to the coast of Novaya Zemlya. Finding his passage to the east blocked, he turned about and regained the coast of the Kola Peninsula where, somewhere to the east of the location of Murmansk, he tarried too long repairing a leaky ship and was frozen in by the ice of an unexpectedly early autumn. The crews, having expected to be in temperate waters by the onset of cold weather, had neither food, adequate clothing nor heated shelter from the Arctic winter on this treeless coast. Trapped in an environment darker and colder than they could have imagined possible, the entire comple-ment of seventy men died of freezing, starvation and perhaps carbon monoxide poisoning in attempts to warm their unheated quarters. The ships and the frozen bodies of their crews were found the fol-lowing summer by Pomor fishers.

Meanwhile, the *Edward Bonaventure* had coasted southeastwards, following Ottar's route from several centuries before, and by freeze-up had attained the Pomor settlement of Kholmagori at the mouth of the Dvina River. From here Chancellor travelled overland by sled to the Moscow of Ivan the Terrible, where he was entertained in grand style and granted the right to trade. The following spring he returned to his ship and sailed to England, having established a pro-ductive commercial route that bypassed the Hanseatic monopoly on Russian trade through the ports of the Baltic. The following year Chancellor retraced his route to Moscow, and was returning to England with Russia's first ambassador when he was shipwrecked and drowned off the coast of Scotland.

The opening of the White Sea route to Moscow diverted most of the Muscovy Company's efforts into establishing a lucrative trade in Russian furs, timber and other products. Richard Chancellor's successor, Anthony Jenkinson, concentrated on exploring over-land routes from Russia to the Orient, and made excursions as far as Bukhara in Central Asia and Qazvin in Persia. Only a remnant effort was maintained in searching for a sea-passage to the northeast. In 1556 Stephen Borough, who had sailed with Chancellor in 1553, set out in the tiny pinnace *Searchthrift* to extend that earlier exploration. With the guidance of local Pomors he found his way along the coast and through the narrow strait to the south of Novaya Zemlya, but was turned back by the stormy and ice-choked waters of the Kara Sea to the east. Borough's report was so pessimistic that it was a quar-ter-century before another English ship sailed this way. In 1580 the Company sent Arthur Pet in the *George* and Charles Jackman (who had recently sailed with Frobisher to Baffin Island) in the *William*, to extend Borough's explorations. They were again turned back by the dismal prospects of entering the Kara Sea, and Jackman and his crew were lost on the return voyage. This loss brought an effective end to the English search for a Northeast passage.

With the English turning their attention to the northern regions of America, Dutch merchants took up the search for new routes to the wealth of the Far East. This venture is associated overwhelmingly with the name of Willem Barents, who made three voyages between 1594 and 1596, on the last of which he died. Although he discovered Spitsbergen, Barents made little further headway to the east than had his English predecessors. His final voyage, however, resulted in a startling and long-lasting contribution to northern knowledge: the first successful Arctic wintering by a European crew, and a published description of that dreadful episode in a narrative that has haunted generations of readers.

Barents made his first Arctic voyage as captain of the *Mercury*, a small vessel in a fleet of three ships outfitted by Amsterdam merchants to penetrate the seas to the east of Novaya Zemlya. The other ships, skippered by captains Nai and Tetgales, sailed through the strait to the south of the island and explored the western portion of the Kara Sea, sighting the Yamal Peninsula and concluding mistakenly that they had passed the mouth of the Ob River. Barents took a more northerly route and attempted to round the northern tip of Novaya Zemlya, but was turned back by ice. The following year seven ships set out to further the exploration of the more southerly route, and Barents served as chief pilot for the expedition. They met ice that was considerably heavier than that of the previous year, and failed to penetrate the strait leading to the Kara Sea. This obstruction persuaded Barents that the coastal route was impracticable, but he remained convinced that a more northerly route to the east could be found through an ice-free polar ocean. Pursuing this belief, in 1596 Barents piloted two ships directly north from the coast of Norway, and almost 500 kilometres offshore chanced upon a tiny and precipitous island; here they killed a polar bear and named their discovery Bear Island. Continuing northward in thick weather they encountered a land of pointed mountains, for which they named the country Spitsbergen, and at the northern end of this land they took an observation at approximately 80° latitude. They were almost certainly the first Europeans to penetrate to this high latitude, but here they were turned back by the edge of the polar ice pack. They retreated to Bear Island, where one of the ships turned for home. The other, commanded by Jacob van Heemskerck, with Barents as pilot and a man named Gerrit de Veer as mate, set off eastwards with the intent of passing to the north of Novaya Zemlya. This was successfully accomplished, but late in the season they became surrounded by the heavy pack-ice of the Kara Sea. They were unable to escape, and in late August took shelter in a shallow bay that they named Ice Haven. Here they were to pass the winter in an ordeal that was painstakingly recorded by de Veer, whose journal was widely translated and published. It became a lasting memorial to Barents and his crew.

Their anchorage provided little protection from a late August snowstorm that drove heavy ice against the ship so that its timbers shrieked "which was most fearfull both to see and heare, and made all the haire of our heads to rise upright with feare." By the first of September the ice froze solid and the explorers began to realize that they had been sentenced to an unimaginable winter on this bleak coast. Expecting the ship to sink as soon as the ice eased its pressure, they salvaged what they could and moved ashore. From driftwood they constructed a makeshift building about eight metres long by six metres wide, with a central chimney and a porch on one end to store firewood and equipment. They named their shelter "het Behouden Huys," the safe house. In 1992 archaeologist Louwrens Hacquebord, with a team of Dutch and Russian archaeologists, found the remains of the house, much diminished from the activities of previous souvenir hunters, but was able to confirm earlier reports that the lower walls had been built log-cabin style with large driftwood logs hauled from a beach six kilometres away. The upper walls and roof had been built of planks salvaged from the ship, covered with wooden shingles. Haquebord's photographs eloquently paint the locality where this first High Arctic wintering took place: a flat and featureless plain of grey-brown gravel.

As winter set in the crew worked to complete their shelter, collect driftwood to feed the fire and furnish their quarters with

Drawing of the structure in which Willem Barents and his crew spent the winter of 1596 on the ice-locked coast of Novaya Zemlya. One man shoots a polar bear from the door of "het Behouden Huys" while others butcher a bear carcass with an axe. The ramp-like structures surrounding the camp are deadfall traps for fox or, judging by their size, for bears. (from S.P. l'Honoré Naber, *Reizen van Willem Barents 1594-1597*, Gravenhage, Martinus Nijhoff, 1917)

sleeping-bunks, a chiming Dutch clock, a lamp burning bear fat and a bath made from a wine cask. While erecting the house they followed the traditional builders' practice of holding nails in their mouths, whereby they were quickly taught the painful lesson learned by every Canadian and Russian school-child: that cold iron burns and tears moist skin. Their ration was reduced to half a pound of bread per day (about half the British naval ration of the time), and they were reduced to drinking water when their beer froze and burst its casks; de Veer notes wryly that they did not have to cool the water with ice or snow. In the twilight and the blowing snow they were continually stalked by bears that had never before seen humans, some of which were bold and curious enough to board the ship and try to enter the house. Many were killed, and their fat provided the lamp-fuel that supplied light through the winter night.

One bear was erected as a gruesome frozen statue that gradually disappeared beneath the deepening snow. Foxes swarmed around the house, stealing anything they could; an early engraving shows deadfall traps surrounding the house, and the numbers of fox bones that Hacquebord found at the site indicate that the tiny animals were a significant food source for the marooned men. In November the cold of full winter descended, and they learned that their uninsulated shelter provided little protection. Snow fell continuously, and huge drifts blocked the doors. Frost penetrated the flimsy walls, and froze everything that wasn't next to the fire in the central hearth. Men burned their clothing trying to get warm while the other side of their body froze. Their woollen clothes were not adequate for Arctic cold, and froze like iron when they were wet; the foxes came to the rescue, as only crude garments sewn from fox-pelts saved them from freezing. On the night of December 7 they treated themselves to a large fire using sea-coal carried from. the ship, and as it died down they stopped up the chimney and every other crack in order to preserve the heat. They went to bed "well comforted with the heat, and so lay a great while talking together." Eventually they were overcome with dizziness, and most were close to being unconscious before they had the wit to open the door, flooding the house with cold air and saving the entire crew from certain death by carbon monoxide poisoning.

Christmas day passed with foul weather and the tormenting sound of foxes running over the roof of the house, yet all but one of the seventeen men were still alive. Twelfth Night was celebrated with pancakes and a single biscuit sopped in the last of the wine: "And so supposing that we were in our owne country and amongst our friends, it comforted us as well as if we had made a great banquet in our owne house." Scurvy appeared, but may have been held at bay by the amount of fresh fox and bear meat in their diets. They were poisoned by eating bear liver, which contains toxic amounts of Vitamin A, and some almost died. Their health improved with the return of the sun at the end of January, after which the officers ensured that everyone exercised outdoors, hauling driftwood from the distant beach and even improvising a game of golf among the wandering bears on the frozen plain. Their ship was damaged beyond repair, so two boats had their gunwales raised with extra planks and were loaded with supplies for an eventual escape. On June 13 the ice opened enough to launch the boats, and the refugees worked their way through the pack around the northern tip of Novaya Zemlya and into the Barents Sea. A week later, as they coped with a storm in the midst of the ice, Barents died of scurvy, likely exacerbated by exhaustion. He and a shipmate who died on the same day were taken ashore and hastily buried. The remainder of the crew sailed and rowed southwards along the coast of Novaya Zemlya, crossed the strait to Vaygach Island and then to the mainland. They continued westward to the coast of the Kola Peninsula where they were rescued by Jan Ryp in the ship that they had departed from the previous summer at Spitsbergen. Miraculously, most of the crew had survived their dreadful winter and the remarkable sea voyage that followed. Willem Barents was one of the few left behind on the coast of what was later to be named the Barents Sea, as a memorial to the greatest Arctic explorer of the time. Archaeologists who searched the area during the 1990s were unable to find his grave, and concluded that either he had been buried only in snow or that his grave, hastily scooped into a gravel beach, had been eroded by storms.

De Veer's widely read description of the winter on Novaya Zemlya put an end to European interest in seeking a maritime route to the north of Eurasia. The next phase of exploration was in the hands of the Pomors whose maritime way of life had developed along the White Sea coast. This effort was associated with the late sixteenth- and early seventeenth-century Russian expansion across the interior regions of Siberia, driven by Cossack military prowess in the service of an astoundingly profitable trade in sables and other furs. As the Cossacks advanced eastward they realized that the natural transportation routes of Siberia extend from south to north, along the series of great rivers that rise in the mountains of central Asia and flow northwards through thousands of kilometres of steppe and forest before emptying into the frozen Arctic sea. These rivers are navigable for great distances, and were the most obvious routes for transporting people, freight and messages across the vast Siberian landscape.

Cossacks adapted the river-boat technology that they knew from the grassland valleys of the Don and the Volga to these Siberian rivers, but needed the knowledge and the skills of the Pomors to establish the Arctic coastal links between the mouths of the rivers and with European Russia. By the late sixteenth century, with the upper regions of the Ob River in Russian hands, Pomors had forged a combination of coastal and river routes linking the English and Dutch mercantile bases in the White Sea to the lower courses of the Ob and Yenisey. By about 1620 they were attempting to penetrate much further, as evidenced by the surprising discovery in the mid-twentieth century of the wreck of a small coastal expedition on the bleak eastern coast of Taimyr Peninsula. These venturers had apparently rounded the ice-locked northern extremity of Siberia at Cape Chelyuskin, and had reached almost as far as the mouth of the Lena. Other and more successful expeditions may have completed the voyage without leaving a trace in the archaeological and historical records.

By the 1640s the Cossacks based at Yakutsk on the middle Lena were sailing small vessels to the mouth of the river and along the coast as far as the Kolyma. These ships were known as *kochi*, the term for small clinker-built Pomor vessels with one or two masts, and were probably constructed from hand-sawn pine planks sewn with spruce-root to frames of larch or birch chopped from Siberian forests. In the summer of 1648 one of these Yakutsk Cossacks, Simon Dezhnev, set off on a journey that was one of the most remarkable and least recognized feats in the history of Arctic exploration. In June his fleet of six tiny boats left the mouth of the Kolyma and headed east. They worked their way through the summer shore leads and occasional stretches of open water between the coast and the permanent pack-ice of the East Siberian Sea. For three months they fought their way along the ice-locked coast of Chukotka, and in the stormy but ice-free waters of early autumn rounded a large headland where the coast turned south and west. On an offshore island (probably one of the Diomede Islands in Bering Strait, or perhaps St. Lawrence Island in the Bering Sea) they had a hostile encounter with Eskimos wearing ivory labrets in their cheeks. Storms plagued the small fleet, and all but Dezhnev's boat were wrecked or disappeared and were presumed lost. When the storms dropped, new ice formed around the boat, and in the cold weather of early winter Dezhnev and his crew were wrecked on the barren coast of the Bering Sea somewhere between Anadyr and Kamchatka. For over two months they wandered along the coast and inland until they finally reached their planned destination in the forested and fur-rich valley of the Anadyr River.

Dezhnev and his crew were the first Europeans to pass the eastern tip of Asia, but being fur-traders and not explorers they made no official claim of discovery, and no notice was taken of their accomplishment. Almost eighty years later the historian Friedrich Müller, while exploring the governor's archives in Yakutsk, chanced upon Dezhnev's report to his commanding officer. If not for this accident of detection, the honour of discovering the strait separating Asia and America would have gone a century later to Vitus Bering, for whom the passage is named. Dezhnev's name appears only on the most easterly promontory of Asia, close to the spot where the first encounter between Russians and Eskimos took place. This headland is more commonly known by its English name, East Cape, bestowed by Captain James Cook when he passed this way 130 years after Dezhnev, and only a few months before his death at the hands of Hawaiian warriors. The almost-forgotten voyage of Simon Dezhnev is probably typical of many unheralded explorations undertaken as part of daily business by men whose names are missing from the history books. More often than not it was these men whose rumoured voyages lighted the way for the explorers credited by history with great journeys and important discoveries.

Peter the Great's interest in both maritime enterprise and geographical knowledge lent massive support to Russian exploration efforts during the early eighteenth century. These culminated in the order given to Vitus Bering, a Danish officer in the Russian navy, to travel overland to Kamchatka where he was to build two ships and determine whether Siberia was joined to America. Unaware that Dezhnev had sailed between the continents midway through the previous century, and that Tsar Peter would die during the coming months, Bering set out for Yakutsk in January 1725. With ship-builders, sailors, soldiers and all of the equipment and supplies needed to assemble ships from the Siberian forests, it took the party two years of hideous effort to reach Okhotsk on the Pacific coast. Here, as ordered, they built the *Fortune* and *St. Gabriel,* and in the summer of 1728 sailed northwards along the coast of Kamchatka and Chukotka. A Chukchi who visited the ship in Anadyr Bay assured them that Chukotka was a peninsula, and that its northern coast extended to the mouth of the Kolyma. This information suggested that Siberia was separated from America, and once Bering had passed East Cape and seen the coast trending to the west, he considered his duty to be complete. Rather than risk being caught by winter in a freezing Chukchi Sea, he decided against sailing west to the Kolyma in order to confirm the matter, and returned to Okhotsk. When Bering sailed through the strait that now bears his name he must have been within fifty kilometres of Cape Prince of Wales on the Alaskan coast; unfortunately, however, the weather was so thick that he didn't see the continent to the east. The following summer Bering took the *St Gabriel* eastwards from Kamchatka, where native informants assured him that a large island lay, but he turned for home before reaching the coast of America.

Arriving back in St. Petersburg in 1730, Bering found that his limited exploration had not convinced the Admiralty that Siberia and America were separate continents, and that they continued to believe the two were perhaps connected somewhere to the north of Bering Strait. He planned a much larger project, which would become known as the Great Northern Expedition, a multi-year programme that saw huge complements of explorers and surveyors dispatched to five distinct sections of the Arctic coast. Bering himself set out once again in 1733 for Okhotsk, but without the implicit support of a great Tsar he now found the difficulties in organizing a distant naval expedition vastly magnified. This time it took seven years to construct two vessels, the *St. Peter* and *St. Paul.* In the summer of 1740 these ships set out eastward from the Kamchatka harbour that had been named Petropavlovsk in their honour, were separated in dense fog, and separately reached the mountainous forested coast of southern Alaska. The *St. Paul* under Aleksey Chirikov lost its ship's boats and their crews, probably to the natives of the area, but on the way home discovered the Aleutian Islands. Bering's *St. Peter* attempted to return by the same route but spent weeks in storms and fog before running ashore on a small island of the Komandorski group. Bering and several others of his crew died of scurvy during the succeeding winter, and the survivors escaped the following summer in a boat built from the remains of their ship.

The career of one of Russia's great explorers was at an end, but his travels were not over. In 1991 the site of the winter tragedy was investigated by a team of Russian and Danish archaeologists, and the bones of Bering and five of his men were excavated and returned to Moscow. Here they were studied by forensic scientists, a reconstruction of Bering's appearance was fashioned from the evidence of his skull, and the skeletons were returned to what is now known as Bering Island where they were interred with full Russian military honours.

Among those associated with the Great Northern Expedition are several of the great names of Russian Arctic exploration: Vasily Pronchishchev, the brothers Khariton and Dmitry Laptev, Dmitry Ovtsyn, and Semyon Chelyuskin who in 1742 finally reached the northernmost point of the Taimyr Peninsula, a cape that now bears his name. By the time that the huge enterprise was completed, most of the northern coast of Siberia had been mapped. The explorers had also accumulated a great deal of information on the offshore waters and their heavy cargo of ice, and it had become apparent that the Northeast Passage did not exist as a sea-route that could be traversed by the ships of the period. Not until the age of iron and steam could such a passage again be contemplated.

The first to accomplish the task was Nils Adolph Erik Nordenskiold, a Swede born in Russian Finland, who had accumulated two decades of experience in exploring the waters and ice caps of Svalbard and Greenland. Backed by Swedish and Russian merchants, as well as by the King of Sweden, Nordenskiold equipped the 300-ton ship *Vega* for Arctic service. Steam-driven and massively strengthened against the pressure of ice, the *Vega* set out from Tromsø in the spring of 1878, accompanied by cargo vessels bound for the Yenisei and the Lena. With a fortunate combination of competence, good equipment and excellent sea ice conditions, the small flotilla reached Dickson (named for one of Nordenskiold's backers) at the mouth of the Yenisei by early August. With little difficulty they continued eastward, rounded Cape Chelyuskin in open water, and in late August brought their final escort safely to the mouth of the Lena. By September, the season of maximum open water, they were steaming along the northern coast of Chukotka, but only 200 kilometres from Bering Strait were stopped by encroaching ice. The planned passage was not to be, but the *Vega* sat comfortably in the ice until the following summer and two days after the ice broke up she steamed triumphantly through Bering Strait. The voyage of the *Vega*, made with such apparent ease and with no loss to life or equipment, was anticlimactic after the centuries of effort and defeat that had gone into its accomplishment.

The successful transit of the passage did not bring an end to heroic and tragic endeavours, and the area continued to produce tales that are classics of Arctic writing. One such story is Valerian Ivanovich Albanov's narrative of his escape from the doomed *Saint Anna*, frozen in the Arctic pack to the north of Franz Josef Land. Albanov was the navigator of a Russian expedition skippered by Georgyi Brusilov that set out in August of 1912 in search of new hunting grounds in the polar sea. With a hastily assembled crew of twenty-three including one woman, supplies for only eighteen months, and no doctor or anti-scorbutics, the venture should never have been attempted, and resulted in the deaths of all except Albanov and one companion. A month after setting out the *Saint Anna* was frozen into the pack-ice of the Kara Sea. Eighteen months later, with the crew ravaged by scurvy and food supplies running low, the ship, still frozen in the ice, had drifted over 1,000 kilometres to the north and far from any land or open water. Faced with the inevitability of death by scurvy, starvation or exposure if ice-pressure sank their ship, in the spring of 1914 Albanov and ten others built makeshift sleds and kayaks and set out for the south. For three months they battled all the hazards of the polar pack and nine men died or disappeared. With incredible endurance and good luck Albanov and Alexander Konrad reached Franz Josef Land and located the well-supplied cabins that had been built by an earlier English explorer. Their luck held, and within two weeks a Russian ship returning home from another failed expedition visited the camp and took the marooned men aboard.

While *Saint Anna* was being carried north in the ice to the Barents Sea, a similar tragedy was unfolding at the other extremity of the Northeast Passage, involving a similar combination of thoughtlessness and bad luck. Half of the complement of the explorer Vilhjalmur Stefansson's Canadian Arctic Expedition was aboard the expedition ship *Karluk* when it became frozen in the ice off the North Alaskan coast in August 1913. Six weeks later Stefansson apparently judged the situation to be perilous, but was unwilling to compromise his venture by abandoning the *Karluk* and bringing the entire party safely ashore. Instead he left the ship, accompanied by five of the most competent travellers, and set off on his own sled-based explorations. The ship and remaining crew of twenty-two, including an Inuit woman and her two young daughters, drifted slowly northwestward and away from the Alaskan coast. They were fortunate to have as their captain the Newfoundland skipper Bob Bartlett, who had accompanied Peary towards the North Pole a few years before. When *Karluk* was crushed by the ice in the polar night of January 1914, far to the north of the Siberian coast, Bartlett organized the salvage of supplies to an ice-camp and then led the entire party to unoccupied Wrangel Island 250 kilometres to the south. With one Inuit companion Bartlett crossed to the Siberian mainland, and with the aid of Chukchi inhabitants travelled a further 600 kilometres eastwards to Bering Strait with hopes of organizing the rescue of his crew. Before he could do so, the twelve survivors of the *Karluk* were picked up by another ship, after having spent eight months on the ice and the barren rocks of Wrangel Island.

As Russia industrialized during the late nineteenth and early twentieth centuries, and as Siberia began to produce exportable quantities of minerals, timber and grain, the need for sea transport quickly grew. Ships destined for the ports of the Ob and Yenisei sailed each year from Murmansk and Arkhangelsk on the White Sea, while ships bound for the Lena and Indigirka and Kolyma rounded Bering

Strait from Magadan or Vladivostok at the eastern terminus of the Trans-Siberian Railway. A commercial shipping route that spanned the entire Arctic coast of Siberia was not accomplished until the 1930s, when a combination of Soviet political will and the development of icebreaking ships with sufficient strength and power to deal with the polar pack led to its development. The Northern Sea Route reduces by half the distance between Russian ports on the Atlantic and Pacific coasts, and soon became the lifeline that fed the enterprises and exported the products of much of Siberia. The Soviet Union became the world leader in icebreaker technology, and by the 1980s several hundred ice-strengthened cargo vessels each year traversed a portion of the Northern Sea Route. As Siberian industry regains its footing after the economic and administrative declines of the early post-Soviet era, and in a period when Arctic sea ice appears to be in retreat before a warming climate, the route promises to regain its recent importance.

Among the technological and administrative matters that defined the development of the Northern Sea Route, one incident—the sinking of the *Chelyuskin*—stands out vividly. Built in Denmark as an icebreaking freighter specially designed for Arctic waters, the *Chelyuskin* set out from Murmansk in 1933 on its maiden voyage, commanded by the explorer and scholar Otto Schmidt and carrying 111 scientists and sailors. In February 1934, while transporting equipment to a scientific station on Wrangel Island, she was caught in heavy ice and sank off the northern coast of Chukotka. The entire complement on board, including men, women, and an infant born during the voyage, was marooned on the ice far from land in the howling darkness of the polar night. For two months they survived, while a rescue effort was assembled and their radio reports made headlines around the world. The entire crew was eventually saved by a small team of pilots flying from an ice-strip built and maintained by terrible effort. For their bravery, the government established a new award: Hero of the Soviet Union. The Northeast Passage may have been conquered by technology, but it still required its share of sacrifice, endurance and heroic deeds.

8 MARTIN FROBISHER'S

GOLD MINES

THE HELICOPTER EMERGED FROM A CURTAIN of falling snow and the pilot pointed directly ahead, then eased back the throttle and began a long slanting descent. For the past hour we had thumped southeastwards down the length of Frobisher Bay, the tundra hills of Baffin Island rolling down to gradually disappear as humped islands and rocky shoals in the heaving waters of the inlet. Now the southern shore was falling away into a distant line of glaciated peaks as the bay opened out into the expanse of Davis Strait and the North Atlantic Ocean. Ahead lay a small almost circular patch of earth and rock that seemed to barely emerge from the waves. This was Qallunaat Island—the name means "White Man's Island" in the language of the Inuit. Here, over four centuries ago, Martin Frobisher's crew of 400 men mined for gold and planned to establish a colony. I was about to set foot on an island that still bears the traces of one of the most bizarre episodes in Arctic history, and one that was to have great significance in the destiny of North America.

Martin Frobisher was an unlikely Arctic explorer. Among the gallant and worldly seamen who graced Queen Elizabeth's court— Hawkins and Raleigh, Gilbert and Drake—Frobisher stood out as a graceless, evil-tempered scoundrel. A sometime pirate, possible

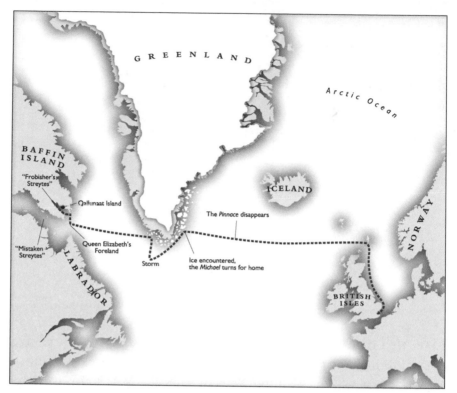

Martin Frobisher's Explorations

slaver on Africa's Guinea coast, suspicious consort of Spanish spies and Irish rebels, he would seem an improbable candidate to undertake England's first bid to discover a Northwest passage to Asia. Yet in three years of voyaging this rough sailor would re-establish the European contact with Arctic North America that had been lost with the disappearance of the Greenland colonies a century before. He would also make the first English claim of possession on the region, lead the first attempt by the English to establish a settlement in the New World, and play a part in the first major gold-mining fraud in North American history. His claim of possession would establish British interest in northern North America, and would be the first step in the eventual establishment of British sovereignty over the northern half of the American continents.

The Frobisher venture grew out of the Arctic experience of the Muscovy Company. The Company had lost interest in developing a trade route to the north of Eurasia after Stephen Borough's 1556 encounter with the ice of the Kara Sea, and had concentrated its efforts on the lucrative commerce with Russia through the White Sea. By the 1570s the Company had developed expertise not only in Arctic voyaging, but in the financing and organizing of such voyages. Michael Lok, a wealthy merchant who was appointed London agent of the Company in 1571, was the obvious man to approach in order to discuss an Arctic venture, and in 1574 such an approach was made. This came from Martin Frobisher, a mariner with whom Lok may have had an old acquaintance stemming from a family venture twenty years before.

Martin Frobisher had been born to a family of local prominence in the Yorkshire Dales, an unlikely beginning for a seaman. Orphaned as a child, he was sent to London in the care of a merchant relative of his mother. Little is known of his education, but those who have attempted to decipher his handwriting and spelling claim that it must have been minimal. His biographer William McFee reports that "He was sent to sea because they could do nothing with him ashore," and goes on to summarize the remainder of his life: "At sea he remained for forty years, with scarcely any rest, and in action he died." Hardship and danger accompanied his career from the beginning. In 1553, at about the age of seventeen, he was one of the few survivors of the heat and fever that decimated the first English expedition to West Africa. The following year he again sailed for West Africa on a venture financed by Thomas Lok and under the command of John Lok, older brothers of Michael Lok. Frobisher volunteered to go ashore as a hostage during trade negotiations on the Guinea coast, and when fighting broke out he was abandoned in the hands of hostile Africans. Three or four years later he reappeared in London, having apparently been presented or sold to Portuguese traders and repatriated by way of Lisbon. He would later report that while in prison in Lisbon he had met an ancient mariner who claimed to have sailed from the Pacific to the Atlantic across the top of North America.

For the following fifteen years Frobisher's name appears only occasionally in historical records, and always in association with trouble: as a successful privateer who repeatedly veered into the piracy of ships from states allied to England, as a possible conspirator with Spanish spies as well as with an Irish rebel imprisoned in London. Then in 1574 he approached Michael Lok with a proposal for an Arctic venture. Perhaps reminded of the old family debt owed to Frobisher for his African abandonment, Lok responded with enthusiasm. Together they obtained a licence from the Muscovy Company, and began to assemble financial backers to undertake a search for a route to the Orient by traversing the seas to the north of the New World. By the spring of 1576 the financing and outfitting were in place, and three tiny ships—*Gabriel, Michael* and an unnamed pinnace, had been purchased or built. The scholar and mystic John Dee was engaged to advise the company on navigational books and instruments, and to teach—unsuccessfully—the complexities of astronomical navigation to Frobisher and his navigator Christopher Hall.

The twentieth-century Arctic explorer Vilhjalmur Stefansson thought that Dee may also have provided more useful information, of a sort that Frobisher could understand: the Norse sailing directions to their Greenland colonies, which he had learned through his scholarly acquaintances on the Continent. When Frobisher's small fleet left the Thames in June 1576—after a gracious leave-taking from Queen Elizabeth who waved from the window of her Greenwich palace—they sailed north to Shetland. Here they turned west to join the old Norse route across the banks south of Iceland, where their small pinnace and her crew of four disappeared in a storm. Three days later they struck a coast with mountains that Hall described as "rising like pinnacles of steeples, and all covered with snow." He measured the latitude as 61°, the location on Greenland's mountainous southeastern coast of the glacier known as Hvitserk (White Shirt) mentioned by the fourteenth-century church official Ivar Bardarson as the Greenlandic landfall for the Norse settlements. Here the remaining two ships were separated amidst the bergs and pack-ice that are carried southwards by the swift East Greenland Current, and the *Michael* took advantage of the separation to turn for home. Left with only the tiny *Gabriel* Frobisher fought free of the ice, barely survived a sudden storm in the Labrador Sea and eventually sighted a bleak headland, which he named after Queen Elizabeth.

Before recounting the itinerary of their explorations in Arctic Canada, it is worth considering how this new world must have appeared to the eyes of Frobisher and his crew. To the Elizabethan mind, the Arctic regions retained some of the magical aura inherited from the ancient and medieval worlds—an immense and bleak Prospero's island where unexpected marvels and hideous dangers might be found. John Dee's studies of abstruse books—the lost *Inventio Fortunatæ* and *Gestæ Arthuri*—provided a fantastic history and geography for the region, as did the bogus Zeno map, which dotted the Arctic with named countries inhabited by strange races. All were incorporated in Mercator's 1569 map, which *Gabriel* carried together with Sir John Mandeville's *Travels*, a fantasy that peopled distant lands with giants, Amazons, cannibals, one-footed races and other peculiar and imaginary populations. The continuous daylight of summer was a wonder that was poorly understood, as was the fact that the Arctic regions attracted the needle of the magnetic compass. On Mercator's map the attraction of the compass was explained by an immense conical mountain of iron protruding from the Arctic sea, while at the North Pole itself stood another mountain so tall that at its summit the sun remained visible throughout the winter.

In the logs and journals published by several participants in the Frobisher venture we find occasional vivid descriptions of a place or an incident that give flashes of insight into the bizarre realm in which these sailors and adventurers found themselves. Finding a dead narwhal on the shore of Baffin Island, they tested its magical powers as an antidote to poison, and declared it to be a unicorn:

> On this West shoare we found a dead fishe floating, whiche had in
> his nose a horne streight and torquet, of lengthe two yardes lacking
> two ynches, being broken in the top, where we might perceive it hollowe,
> into which some of our Saylers putting Spiders, they presently dyed. I
> sawe not the tryall hereof, but it was reported unto me of a trueth: by
> the vertue whereof, we supposed it to be a sea Unicorne.
>
> (Dionyse Settle's account of the 1577 voyage)

The tapered spirals of ivory unicorn horns had long been valued as objects with potent magical powers, their origin in the Greenland trade obscured by tales of mythical horse-like animals from distant eastern countries. Their discovery on Baffin Island confirmed the enchanted quality of this land of ice.

The ice itself was a source of dread and wonder to those who had never encountered it before. Michael Lok's report of the first sight of Greenland typifies the alarm inspired by moving ice:

> ...they had sight of land unknowne to them, for they could not come to set fote theron for the marveilous haboundance of monstrous great ilands of ise which lay dryving all alongst the coast therof... And bearing in nerer to discover the same, they found yt marveilous high, and full of high ragged roks all along by the coast, and some of the ilands of ise were nere yt of such heigth as the clowds hanged about the tops of them, and the byrds that flew about them were owt of sight.

The mariners reacted very differently to the stable sheets of early summer land-fast ice. These brilliant plains veined with azure leads and ponds of fresh water were as delightful to Frobisher's crews as they are to southerners today:

> But now I remember I saw very strange wonders, men walking, running, leaping & floting upon the maine seas 40 miles from any land, without any Shippe or other vessell under them. Also I saw fresh Rivers running amidst the salt Sea a hundred myle from land, which if any man will not beleeve, let him know that many of our company lept out of their Shippe uppon Ilands of Ise, and running there uppe and downe, did shoote at buttes uppon the Ise, and with their Calivers did kill greate Ceales, which use to lye and sleepe upon the Ise, and this Ise melting above at the toppe by reflection of the Sunne, came downe in sundrye streames, whyche uniting togither, made a prettie brooke able to drive a Mill.
>
> (George Best's account of the 1578 voyage)

The brilliance of sunlight reflected from summer ice may have contributed to the mariners' impression of Arctic weather, for their journals often noted the oppressive heat. Although they knew that the winter months would be cold and dark, they appear to have had no concept of the actual savagery of Arctic winter weather. The colony of 100 men that they planned to leave over the winter was to be sheltered in a huge prefabricated wooden barrack, apparently with no insulation and heated only by a Dutch tile-stove. While such a shelter might have been adequate in Orkney or Shetland, it would have provided little protection from a Baffin Island winter. Before leaving for home in 1578 they even tested the agricultural potential of the country:

> *Also here we sowed pease, corne, and other graine, to prove the fruitful-*
> *nesse of the soyle against the next yeare.*
> —George Best

These were the acts of men who had no true idea of the Arctic environment. To Frobisher's crews this treeless countryside resembled the bleaker regions of their homelands: the deforested mountains of Yorkshire, Scotland and the Hebrides. Their response to Arctic conditions was that of people who sought the familiar in the alien, combined with an expectation of encountering magical wonders in this distant and alien part of the world.

In July of 1576, when Frobisher made his landfall at the high cape that he named Queen Elizabeth Foreland, he had only begun to encounter the wonders of the Arctic. The headland was part of Resolution Island, the most easterly outlier of Arctic North America, and the land was so enclosed by ice that it was several days before they could put a boat ashore on a small island. Here Frobisher instructed the crew to collect samples of whatever they found "in token of Christian possession" of the land, and one of the sailors brought back a small black stone that was to be of great consequence to the future history of the venture.

Eventually the ice cleared, and Frobisher sailed *Gabriel* into an inlet that he assumed to be an open passage to the west. George Best's account suggests that the voyagers thought they had encountered a northern counterpart to the dangerous but relatively short Strait of Magellan, which led from the southern Atlantic to the Pacific Ocean:

And that land uppon hys right hande, as hée sayled Westward, he judged to bée the continente of Asia, and there to bée devided from the firme of America, which lyeth uppon the lefte hande overagainst the same. This place he named after his name Frobishers Streytes, lyke as Magellanus at the Southweast end of the worlde, havyng discouvered the passage to the South Sea (where America is devided from the continente of that lande, whiche lyeth under the South Pole) and called the same straites Magellanes streightes.

We now know Frobisher's "Streytes" as Frobisher Bay, a tapering 200-kilometre inlet that terminates in the hilly interior of Baffin Island. Frobisher didn't reach the western end of the bay. After sailing westward for several days, he and Christopher Hall went ashore and climbed a hill to try to find their way through the maze of islands and shoals that practically blocks the inlet. But before they could detect the closed western end of the bay, barely thirty kilometres away, they were distracted by another phenomenon: the Inuit of Baffin Island who arrived prepared to trade with the newcomers.

Over the next few days the English and Inuit traded, ate together and competed at gymnastics in *Gabriel's* rigging. The English understood from their new acquaintances that the open sea lay to the west, and could be reached in two days paddling by kayak; this misunderstanding probably arose when the Inuit tried to tell them that the bay ended at this distance to the west. One man agreed to pilot *Gabriel* through the islands to their western destination, and was taken ashore in the ship's boat to fetch his kayak. This agreement led to the episode that would end the summer's exploration and plague Frobisher's venture over the next two years.

Gabriel's only boat disappeared around a point of land, and—against strict orders—the five sailors manning it appear to have gone ashore with their passenger. Neither they nor the boat were seen again. Assuming that they had been captured by hostile and probably cannibalistic Inuit, but helpless to follow, Frobisher cruised offshore blowing trumpets and firing his cannons. He eventually lured an Inuit man close enough to be captured from his kayak as a hostage, but this produced no exchange. Eventually, with snow falling and

Frobisher's search for a Northwest Passage in 1576 ended amidst this confusion of islands near the head of what is now Frobisher Bay. (Robert McGhee, Canadian Museum of Civilization)

winter approaching, Frobisher abandoned hope of retrieving the men and sailed for home.

Three centuries later the Inuit of Frobisher Bay had a historical tradition concerning five European sailors who had long ago been accidentally or deliberately marooned in the area. The men had been kept alive over the winter by the kindness of the local leader, but perished the following summer while trying to sail away in a boat that they had built. We shall never know the exact story of the five lost sailors, but may guess that the truth lay somewhere between English suspicion and Inuit oral history. Young men who had spent the previous months in the cold cramped quarters of a tossing ship would have found it impossible to refuse a friendly invitation to come ashore, to walk in the sunshine, eat fresh meat, trade trinkets for valuable furs and be introduced to their hosts' shy tattooed daughters and wives. Inuit hospitality, combined with fear of punishment aboard ship, may have kept them ashore longer than planned. Although the camp leader may have been able to ensure the lives and well-being of his guests, he may not have been able to prevent other men from commandeering the wooden boat, which would have been immensely valuable to maritime hunters. Frobisher's kidnapping of one of their number would have brought an end to any possibility of compromise, and the fate of the sailors was sealed. The Inuit hostage fared no better; shortly after reaching England he died "of colde which he had taken at Sea."

Gabriel reached England safely, Frobisher told of having almost traversed the Northwest Passage, and his Inuit hostage was presented as proof that he had reached an Oriental land. He and Lok laid plans for another expedition the following year, which they had no doubt would reach the Pacific and fulfill their dreams of wealth. Other backers were more difficult to persuade, but at some point during the winter interest in the voyage quickened, although its focus shifted significantly. It isn't clear why someone thought to test the black stone that had been picked up from an Arctic beach the previous spring; one report tells that a fragment of it was given to the wife of an investor, who threw it in the fire and noticed that it "glittered with a bright Marquesset of golde." In any case, samples of the stone were provided to several goldfiners and assayers, all but one of whom pronounced it worthless. The final assayer, an alchemist of dubious training and experience, declared that it was high-grade gold ore and would bring vast wealth to those who could mine it in quantity.

With the curious momentum that develops when rumours of gold reach the street, backers for a second voyage were quickly found and Queen Elizabeth herself invested £1,000. In the spring of 1577 Frobisher headed northwest across the Atlantic with 120 sailors and miners aboard three ships: the tiny *Gabriel* and *Michael* of the previous year, plus the much larger *Aid*. Although exploration was still a part of his instructions, Frobisher was to consider it only after he had mined a cargo of ore for the ships. The first weeks of the voyage were disappointing. Although they were able to locate the island from which the black stone had originally come, they could find no more samples, much less a vein of ore worth mining. They captured more Inuit hostages and searched the camp of a party with whom they had a pitched skirmish, but could get no further news of the five sailors

The small flat island at the centre of the photo, emerging from a snowsquall over Baffin Island's Frobisher Bay, was the site of the first English attempt to establish a settlement in the New World, and of the first major mining fraud in North America. (Robert McGhee, Canadian Museum of Civilization)

whom they had lost the previous summer. The letter that Frobisher wrote and sent to them by Inuit couriers is touching in its sincerity, evocative of the religious aura that surrounded the entire enterprise and ineffectual in its combination of pious pleas and threats of genocide:

> *In the name of God, in whom we al beleve, who I trust hath preserved your bodyes and soules amongst these Infidels, I commend me unto you. I will be glad to seeke by all meanes you can devise, for your deliverance, eyther with force, or with any commodities within my Shippes, whiche I will not spare for your sakes, or any thing else I can doe for you. I have aboord, of theyrs, a Man, a Woman, and a Childe, which I am contented to deliver for you, but the man which I carried away from hence the last yeare, is dead in England. Moreover, you may declare unto them, that if they deliver you not, I wyll not leave a manne alive in their Countrey. And thus, if one of you can come to speake with me, they shall have eyther the Man, Woman, or Childe in pawne for you.*

And thus unto God, whome I trust you do serve, in hast I leave you, and to him we will dayly pray for you. This Tuesdaye morning the seaventh of August. Anno. 1577.

Yours to the uttermost of my power
Martin Frobisher

Eventually, while searching the outer coasts of Frobisher Bay for either hostages or gold, the ships came across a small island containing an abundance of black rock resembling the sample that had been assayed as gold ore. The island was named after the Countess of Warwick, the wife of an important backer of the expedition, and the party set about excavating a mine and loading ore.

The mine dug by Frobisher's men was clearly visible four centuries later as our helicopter descended towards Qallunaat Island: the huge notch cut in the sea-cliffs unmistakably identifies Qallunaat with the island that Frobisher named for the Countess of Warwick. The rocks forming one wall of the mine still bear the tool-marks left by miners over 400 years ago, and the back-breaking work carried out here is easily imagined. With picks and heavy hammers, wedges and pry-bars, the men split off blocks of rock and tumbled them to the beach below. At high tide they were loaded into boats and ferried to the waiting ships, hoisted by derrick in huge wicker baskets and stowed in the hold. They had loaded almost 200 tons of ore in twenty days, and by late August the men were exhausted, the equipment was broken, and winter was fast closing in. George Best reported that

Thurseday the 22 of August, we plucked downe our tentes, and every man hasted homewarde, and making bonefires uppon the toppe of the highest Mount of the Iland, and marching with Auntiente displayd round aboute the Iland, we gave a vollie of shotte for a farewell, in honour of the right Honourable Lady Anne, Countesse of Warwicke, whose name it beareth; and so departed aboorde.

Standing on the island's summit in a late August snowfall, one can see nothing but bleak rocks, cold water and the distant icefields rising above the southern shore of the bay. It is easy to imagine the

The marks of tools used by Frobisher's miners remain clearly visible on the rock wall of one of the Qallunaat Island mines. (Robert McGhee, Canadian Museum of Civilization)

small ceremony that took place here in 1577: the flaring bonfire, the procession of tired and limping men in tattered clothing following the flag around the rocky island, the curses and prayers of thanks mingling with the salute of gunfire. One can sense the relief with which the Englishmen turned their backs on this alien place of discomfort and constant toil, and welcomed the cramped quarters of ships that would soon be pitching southwards to warmer latitudes and distant homes. Their relief was more than matched by the terror of the Inuit hostages—a man, woman and her infant child—as they saw their only known world slipping out of sight.

By the following winter all three Inuit captives had died, and rival assayers were at work estimating the amount of gold contained in the ore that Frobisher had brought home. The investors found themselves repeatedly called upon to provide additional capital—to pay off the ships' crews, to build smelting furnaces, to finance another expedition—or forfeit their shares in the enterprise. By the early spring of 1578 Frobisher became so frustrated with the assayer Jonas Schutz's failure to produce results that would attract investment that, according to Michael Lok's account "He drew his dagger and furiouslye ranne uppon Jonas, beinge in his worke at Tower hill, and threatned to kill him yf he did not finishe his worke owt of hand, that he might

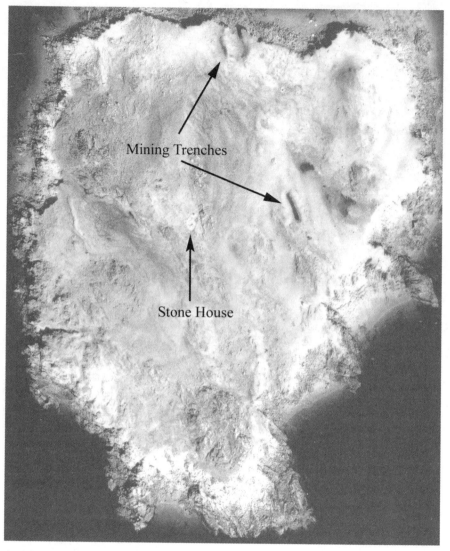

Aerial photograph of Qallunaat Island. The remains of two large mining trenches are clearly visible, as are the remains of the small stone house built at the summit of the island. (National Air Photo Library, Ottawa, Canada)

be sett owt againe on the thirde voiage." The assay that Schutz finally completed showed negligible value, but by this time a full-fledged gold play was developing and a mere assay could easily be ignored.

In June of 1578 the largest fleet ever assembled for an Arctic expedition set out for the supposed gold mines of Baffin Island.

Fifteen ships carried 400 mariners, soldiers and miners on a voyage that would not only gather the wealth of the country through mining and trade, but also establish a settlement that would serve as a mark of English possession in this region of the world.

The fleet made landfall in southwestern Greenland, where Frobisher went ashore and examined dwellings from which the Inuit occupants had fled. Noting their similarities to the people who occupied the shores of Frobisher's Straits, he concluded that the countries were joined and claimed Greenland under the name West England. They then set out to cross Davis Strait, but were soon beset not only by fog and storm but by a vast herd of whales; one was struck by a ship under full sail, bringing it to an abrupt halt, while the whale "made a great and ugly noise, and caste up his body and tayle, and so went under water." When they reached the western side of Davis Strait the ships encountered heavy ice, from which most escaped in great danger when a storm sprang up. The fleet was scattered and was not to be reunited for several weeks. Frobisher led a small convoy westward into what he adamantly insisted was Frobisher's Straits but which eventually turned out to be Hudson Strait, the wide and turbulent channel between Baffin Island and mainland North America and the route to the inland sea of Hudson Bay. Frobisher eventually named it Mistaken Strait as he turned about to retrace his route, but his mistake was to have enormous consequence as later explorers followed it into the heart of the continent.

By the beginning of August, after two months of hardship and danger, most of the ships had assembled off the Countess of Warwick Island. Here the first English service of Thanksgiving in the New World was held, in which the minister Robert Wolfall

> ... *made unto them a godly Sermon, exhorting them especially to be thankefull to God for theyr strange and miraculous deliverance in those so dangerous places, and putting them in mynde of the uncertainetie of mans life, willed them to make themselves alwayes ready, as resolute men, to enjoy and accept thankefully whatsoever adventure his devine Providence should appoynt.*

The miners were then set to work excavating another mine on the Countess of Warwick Island, as well as optimistically named mines at other locations where similar black rock was found: Fenton's Fortune, Best's Blessing, Winter's Furnace and the Countess of Sussex Mine. An assaying shop was set up on the Countess of Warwick Island, which continued to serve as the headquarters for the expedition, and the ore coming in from other locations was tested for value. A meeting was held to discuss the plans for establishing a wintering-party of 100 men, since, very fortunately for those who were to have been left, half of their prefabricated barracks had been in a ship that had been sunk by the earlier storm in the ice. Much of their supplies had also been lost or spoiled, including most of the beer, which was a staple food. It was therefore decided that no colony would be established that year, and we can imagine the relief with which the news was greeted by the erstwhile colonists. It would be almost a decade before the English made their next attempt to establish a settlement in the New World, at Roanoke in what is now North Carolina.

By late August over 1,200 tons of rock had been quarried from the various mines and stowed in the ships. The final act of the expedition was to build a small stone house, as an experiment to see how English buildings survived Arctic winters, and as a means of attracting the Inuit to peaceful trade. This house, the first to be constructed by the English in the New World, was hastily erected at the summit of the Countess of Warwick Island, and its walls hung with mirrors, bells, whistles, toy figures and other items. When it was completed, the crews listened to a sermon, celebrated communion and immediately set out for home in a heavy storm. Despite a near-panic in attempting to get clear of the dangerous lee shore before the storm set in, all of the miners were brought aboard and the ships made a safe Atlantic crossing, although one of the fifteen was lost on the coast of Ireland. Forty men died on the expedition, most of them on the journey home as injuries, malnutrition, scurvy and exposure took their toll.

Today Qallunaat Island looks much as it must have appeared a few weeks after Frobisher left and after the local Inuit, who had been

The tumbled remains of the first stone house built by the English in the New World lie at the summit of Qallunaat Island. (James Tuck)

closely watching the visitors, demolished the standing structures for their wood and nails. The mines remain as two great scars on the island's surface; beside one of them lies a small heap of black rocks abandoned before they could be hauled away. Two rectangular arrangements of boulders mark the foundations of small assaying shops, and are surrounded by scatters of charcoal, slag and fragments of crucibles. Indistinct scatters of boulders probably mark the locations of tent camps where the miners slept on mattresses of cold gravel. Two upright boulders in a field of gravel may be makeshift headstones marking miners' graves. Excavations in one of the mines produced fragments of ceramic tiles from a Dutch stove and barrels of dried peas, biscuit and other supplies for the erstwhile colony, which had been buried here for future use. A tumbled heap of boulders and plaster fragments at the summit of the island marks the location of the stone house. This is still very clearly Frobisher's Countess of Warwick Island, and the remains of an Elizabethan settlement are still a clearly visible part of the terrain. The sense of an historical landscape, of a place where the past is very close, pervades the atmosphere. On a quiet walk around the island one is conscious of the same unchanged vistas seen by Frobisher and his men, the same rocks and patches of gravel underfoot, the same constant, bitter wind. The small island is easily populated in the imagination with hundreds of toiling, wretchedly uncomfortable and hopelessly optimistic men.

The final episodes of the Frobisher venture did not occur here on Qallunaat Island, but were played out in England over the following year. Michael Lok and other backers of the venture had constructed the largest smelting complex in England in order to extract gold from the ore, and hundreds of tons were transported to its location in the small Thames Valley town of Dartford. Trains of horse-drawn carts carried over 1,500 loads of ore to Dartford, where over the following winter it gradually became apparent that the utmost efforts and ingenuity could not extract an ounce of gold. Most of the backers lost their investment, the company disbanded in a welter of lawsuits and Michael Lok ended his long and honourable career in debtor's prison. Martin Frobisher went back to sea on occasional assignments from the Queen, and probably more often in his old business as a pirate. His reputation was recovered only by the heroic part that he played in the battle against the Spanish Armada, and a few years later he was killed while leading an attack on a Spanish fort in Brittany. This final service to his Queen dissolved any blame for his role in the disastrous affair of the Northwest Passage and its imaginary gold mines.

The entire story of the deception behind this first major gold-mining fraud in Canadian history has never been discovered. The black rock excavated with such labour and transported across the Atlantic was not fool's gold, as is commonly believed, but several varieties of dark metamorphic rocks containing abundant mica, which sparkles in the sun. While carrying out an archaeological investigation of the remains of Frobisher's activities on Qallunaat Island, one evening we heated a small chunk of this rock over the flame of our camp-stove. With intense heat the mica oxidized to a golden-brown colour that might possibly be mistaken, but only by an eager and willing mind, for evidence of gold. Any competent assayer

The blocks of black stone in this wall around the Queen's Manor House in the Thames Valley town of Dartford were mined from Baffin Island and carried here to be smelted for gold. (Robert McGhee, Canadian Museum of Civilization)

would soon prove the rock to be worthless, but Elizabethan assaying was an art not far removed from alchemy, and its results were often unpredictable and difficult to interpret.

The final judgement of Martin Frobisher's accomplishments as an Arctic explorer must acknowledge that he did not discover a Northwest Passage to Asia; that his purported gold mines turned out to be worthless; and that his attempt to plant the first English colony in the New World was a failure. The mines were soon forgotten and even their location was lost to history, aside from the oral tradition of the Baffin Island Inuit, who preserved the stories of old events associated with the place that they called White Man's Island. Nevertheless, the Frobisher venture had a major and largely unintended consequence: it was the faltering first step in a series of events that eventually led to English sovereignty over the northern portions of the New World.

To explain this claim we must return to John Dee, the mystical cosmographer and promoter of English ventures in the Arctic, and the man who attempted to teach navigation to Martin Frobisher. Dee is credited with having first conceived the idea of a British Empire, and Frobisher's land-taking in Arctic Canada was the immediate stimulus for the idea. After Frobisher's return from the final 1578 voyage, Queen Elizabeth asked Dee to produce an argument proving her ownership of the Arctic countries that Frobisher had discovered and claimed. Dee's response is probably the large parchment document that he presented to the Queen in 1580, and that is today housed in the British Library. On one side is a map of the world, and on the other a detailed justification based on historical precedent—from the mythical Arctic explorations of King Arthur to those of Martin Frobisher—substantiating Elizabeth's sovereignty over the northern portions of America and the islands lying to the north of it.

Dee's hand can be seen in the patent issued to Humphrey Gilbert for exploration of North America, and under this patent Gilbert's brother Adrian granted Dee in 1580 the rights to all New World lands to the north of 50° latitude, which would encompass most of what is now Canada. Dee joined with Adrian Gilbert and John Davis as "The colleagues of the fellowship of the New Navigations Atlantical and Septentrional" in applying for a charter to explore and hold a trade monopoly in these regions. He also instructed Davis in navigation and, proving to be a far better pupil than Frobisher, between 1585 and 1587 Davis undertook three well-conducted expeditions that produced a basic survey of the coasts between Greenland and Labrador. Most important for further exploration, he described and determined the correct latitude of Frobisher's Mistaken Strait, the "very great gulfe" that lay off the northern tip of Labrador and was the passage to the great inland sea at the heart of central Canada. By 1610 Henry Hudson had explored that body of water, and within a century of Frobisher's voyages the English had established settlements at the bottom of Hudson Bay. The fur trade that was established from these posts extended far up the rivers of the Canadian prairies, bypassing the trading empire of the French in the St. Lawrence and Mississippi valleys. The political, social and linguistic character of northern North America is in large part an incidental result of Martin Frobisher's unsuccessful search for gold, and for an Arctic passage to the wealth of Asia.

9 THE RAPE OF SPITSBERGEN

THE ARCTIC OCEAN IS THE LEAST PRODUCTIVE SEA on earth. Cold, deep and sealed with ice that blocks sunlight for six to twelve months of the year, it supports only minimal biological activity. In exceptional locations, however, circumstances allow pockets of productivity to develop, and these dense concentrations of life are the more remarkable for their contrast with the surrounding emptiness. One such concentration lies far to the north of Europe, midway between the northern coast of Norway and the Pole.

This florescence of Arctic life has its origin far to the south, in the tropical waters that feed the current known as the North Atlantic Gyre. Driven by trade winds and the forces of the turning earth, a vast vein of blood-warm water sweeps west from Africa to the Caribbean, circles the Sargasso Sea and swings northward along the coast of North America. Here it is known as the Gulf Stream, 100 kilometres wide and a kilometre deep, moving at a rate of more than 100 kilometres a day. When it curves eastward from the New England coast and crosses the Grand Banks south of Newfoundland, its fog-haunted boundary with the subarctic water of the Labrador Sea is marked by a temperature difference of between 10° and 15° C. Icebergs drifting south from Greenland melt like ice-cubes in a warm bath. Renamed

the North Atlantic Drift, the current streams northeastward between
Scotland and Iceland, warming the overlying air to an extent that
allows palm trees to grow in the former and birch trees in the latter.
One branch of the current washes the Norwegian coast while
another swings offshore and strikes northwards across the Norwegian
Sea. Here the still-warm and salty tropical water collides with the
southward-drifting ice of the polar pack in the vicinity of the group
of islands once known as Spitsbergen and now more commonly as
Svalbard. As it cools, the salty tropical water becomes denser than the
cold Arctic water. Near Svalbard it sinks and turns southward to flow
through the oceanic depths, to eventually replace the water sent north
by the surface currents of the gyre.

Ocean currents do not simply slide past one another as they do on
charts, or slip quietly into the depths as they give up the last of their
warmth. They swirl and eddy in chaotic fashion, swaying beneath the
force of surface winds, and drawing up occasional spouts of deep
water to replace that sinking from the surface. The resulting chaos
provides a medium that combines the elements required for abundant
life. The warm and salty southern water gives up heat to the cold,
oxygen-rich and relatively fresh water of the Arctic; surges of
nutrient-rich deep water are displaced upwards as the salty water
sinks; pack-ice melts back, releasing its concentrated blooms of early-
summer algae and allowing the continuous sunlight of summer
months to penetrate the upper layers of the sea. Plankton blooms and
dies, shrimps and krill, squid and small Arctic cod multiply into vast
shoals. The shallow sea-bed overlying the continental shelf is
encrusted with shellfish and a variety of other creatures, generously
fed by the detritus of the animals living and dying in the waters above.

At the top of the food-chain are the sea mammals. Immense herds
of walrus loll on the rocky shores and ice floes after feasting on the
sea-bed buffet. Colonies of harp and hooded seals mingle with the less
gregarious ringed and bearded seals, and with flashing pods of white
beluga and single-tusked narwhals. Everywhere are the spouts of large
whales—sei, minke, blue, humpback, right and bowhead—calmly and
resolutely cruising the upper levels of the sea, filtering the abundant
soup of tiny living forms through the curtains of baleen hanging from

their upper jaws. Above them swings a whirling canopy of seabirds alert for remnants of the constant feeding below, or hunting for themselves among the small creatures of the upper water. White bears and black orcas complete the picture, elegant killers patrolling the ice and dark water to prey on their fellow sea mammals.

This is not a picture of Svalbard today, but of what it was four centuries ago. For together with the fortuitous confluence of ice and currents that supplied the nutrients for plankton and other small creatures, life in and around the archipelago was dependent on another fortunate circumstance: isolation from the lands inhabited by humans. Located almost midway between Norway and Greenland, the islands were among the last discovered by mankind. Neolithic hunters exploring the icefields northwards from the Eurasian coast had long ago reached the nearer islands as they followed migrating reindeer to Novaya Zemlya, Severnaya Zemlya and the New Siberian Islands, all of which lie within a few hours' or days' walk of the mainland. By about 4,000 years ago they had reached Wrangel Island lying over 100 kilometres off the north coast of Chukotka, and had wiped out the mammoths that until then had survived in that isolated locality. At about the same time, their relatives known as the Tuniit spread eastward across the North American Arctic and around the coast of Greenland. Only Svalbard, the adjacent Franz Josef archipelago to the east, and a few of the most isolated islands of the Canadian Arctic archipelago, were not discovered and used by these ancient explorers and hunters of the Arctic.

Svalbard lay far to the north of the Viking exploration route across the North Atlantic. Icelandic annals record a storm-driven ship that, in the late twelfth century, discovered a land called Svalbard (the name means "cold coast") but it seems likely that the name refers to the northeastern coast of Greenland, or perhaps tiny Jan Mayen Island, 400 kilometres to the north of Iceland. Almost certainly the first humans to sight the archipelago were the crew of the 1596 Dutch expedition piloted by Willem Barents in search of a sea route to Asia. Sailing northwards in hopes of penetrating an ice-free polar sea, Barents's ships first encountered Bear Island, the small southern outlier of the archipelago. Continuing northwards in foggy weather they encountered a coast backed by pointed mountains, for which they named the land Spitsbergen. The crew went ashore and gathered thousands of birds eggs, including those of barnacle geese, and were probably the first Europeans to see the nesting grounds of these small geese, which migrate northwards across Europe each spring. Barents's mate, Gerrit de Veer, noted the widespread medieval myth that barnacle geese grew on trees, "but this is now found to be contrary, and it is not to be wondered at that no man could tell where they lay their egges, for no man that ever we knew had ever beene under 80 degrees." Prevented by ice from circling the northern end of the land, Barents sailed southeastwards to Novaya Zemlya where, after the first heroic European attempt to pass the winter on an Arctic island, he died.

The archipelago discovered by Barents lies between 74° and 81° North, 10° and 35° East, at the confluence of the Norwegian Sea, the Barents Sea and the Arctic Ocean. A tight cluster of five main islands with a few small outliers, Svalbard has an area somewhat smaller than that of Ireland or Hokkaido. In contrast to those lush islands, this is a landscape of steep and barren mountains rising abruptly from the coast, snow-covered for much of the year and supporting no vegetation larger than ankle-high shrubs of Arctic willow. The large northeastern island, Nordaustlandet, is almost entirely buried beneath a permanent ice-sheet, and glaciers cover significant portions of other islands. Winter sea ice locks the archipelago for six or more months of the year, and rarely retreats far from the northern and eastern coasts during the summer melt. Only the island that is today known as Spitsbergen, the largest and most westerly island that is washed by the warm current of the North Atlantic Drift, is relatively free of snow and ice during the summer months.

Eight years after Barents's discovery of Spitsbergen, an English ship, the Muscovy Company's *Speed*, again touched at Bear Island where they encountered massive herds of "sea-horses"—walrus—resting on the beaches. Although walrus had long been hunted by the aboriginal peoples of the circumpolar world, and their ivory tusks had formed a significant part of Greenlandic Norse trade with medieval Europe, the animals were new to most European sailors.

The pointed mountains that gave rise to the name Spitsbergen (Robert McGhee, Canadian Museum of Civilization)

The Atlantic walrus is a huge beast, the bulls averaging three-quarters of a tonne and cows over half a tonne, encased in a bulky layer of insulating blubber and a thick, tough hide. Ashore they are ungainly creatures that move ponderously and with apparent effort, but when attacked they take to the water where they are capable of swift movement and extensive dives. Feeding on molluscs and other seabed creatures, they apparently use the sensitive bristles on their snout to detect their prey. This appears to be an efficient means of acquiring a living, for relatively little time is spent feeding compared to the hours and days passed in large companionable herds either ashore or on the ice. Calves are born on spring ice and soon take to the water, remaining with their mothers for two years, during which time the mother suckles and protects the calf. Orphaned calves are often adopted by other females, and the entire herd is fiercely protective of its calves and even its adult members. The shallow and productive seabed surrounding Svalbard, together with the proximity of ice and rocky shores where they were entirely free from predators, made the archipelago an ideal home for walrus. Biologists estimate that some 25,000 animals lived on the coasts of Spitsbergen when human hunters first arrived.

Stephen Bennet, master of the Muscovy Company's *Speed* in 1604, described how his crew experimented in hunting this strange and potentially dangerous prey:

> *It seemed very strange to us to see such a multitude of monsters of the sea lye like hogges upon heapes: in the end wee shot at them, not knowing whether they could runne swiftly or seize upon us or no…. Some, when they were wounded in the flesh, would but looke up and lye downe againe. Some were killed with the first shot; and some would goe into the sea with five or sixe shot; they are of such an incredible strength. When all our shot and powder was spent, wee would blow their eyes out with a little pease shot, and then come on the blind side of them, and with our carpenter's axe cleave their heads. But for all we coulde doe, of above a thousand we killed but fifteene.*

The following year Bennet's crew learned to kill walrus with lances, and rendered 11 tonnes of oil. In 1606 they had become so expert that they killed between 600 and 700 walrus in six hours, rendered 22 tonnes of oil and filled three hogsheads with tusks.

The activity of such hunting crews is typified by the journal entry for June 5, 1610, made by Jonas Poole who was master of the Muscovy Company's ship *Amitie* at Bear Island. The day began when 200 walrus were reported on a rock:

> *I tooke both the boat and the skiffe, with all my company and went to the rock, and in going thither I sleue a bear; but when I came to the rocke, the beasts begun to goe into the sea, then I presently went on land, with all my company, and slue eightie beasts, whose teeth I tooke, and in going aboord slue another bear.*

The returns of the day were increased by salvaging a narwhal tusk and a quantity of baleen from a nearby beach. During their brief voyage the crew of the *Amitie* killed 120 walrus, 51 reindeer and 30 bears as well as capturing two bear cubs alive.

Despite the value of ivory, baleen, bearskins and walrus hides, the major profit to be made in such voyages was in oil. Large quantities of oil were required by European societies as ingredients for soaps and medicines, material used in the manufacture of cloth and leather goods, lubrication for wooden machinery and fuel for lamps. Olive oil had traditionally supplied the needs of southern countries, but vegetable oils and oils rendered from domestic animals were clearly insufficient for the growing societies of northern and western Europe. Sea mammal oil was an excellent alternative, and since late medieval times had been a source of wealth for the Basque communities that had developed the skills and techniques required for hunting the large whales of the Bay of Biscay. Having exhausted this source, Basque whalers turned in the early sixteenth century to the newly discovered whaling grounds in Labrador and Newfoundland waters. By the 1580s these whale populations too were declining rapidly, and by the early seventeenth century sea mammal oil was an expensive commodity in a Europe that had come to depend on the substance.

The walrus and even the bears killed by the *Amitie* in 1610 were rendered for oil, but Jonas Poole also reported "great store of whales" in the waters off Bear Island. Three years earlier Henry Hudson had noted the same when, searching for an open-water route to Asia, he had reached the west coast of Spitsbergen and entered a fiord that he named Whales Bay (now Kongsfjord). In 1611 the Muscovy Company outfitted an experimental whaling voyage, including in the crew six Basque whalers who killed the first whale taken in Spitsbergen waters. The English soon learned the Basque hunting technique, which involved a barbed iron harpoon attached to a rope. The weapon was thrown from a small rowing-boat and then tied to the bow of the boat, which became a drag to slow the wounded whale until it could be killed with lances. This hunting technique was especially effective against the bowhead whale (*Balaena mysticetus*) of the Arctic seas, a close relative to the right whales that the Basques had hunted in home waters and off Labrador. This placid and slow-swimming animal could be approached by men in a small boat, and was insulated by so much blubber that it floated when killed, so that it could be towed ashore. It has been estimated that more than 40,000 bowhead whales roamed the waters around Spitsbergen at the beginning of the seventeenth century.

The remains of three furnaces that once supported huge kettles for rendering whale oil, on the coast of Edgeøya, Svalbard. (Robert McGhee, Canadian Museum of Civilization)

In 1612 two ships took seventeen whales as well as many walrus, and rendered 180 tons of oil. Jonas Poole reports of this voyage that "the whales lay so thicke about the ship that some ran against our cables, some against the ship, and one against the rudder. One lay under our beake-head and slept there a long while." The profitable returns of this voyage set off an avalanche that would practically exterminate the Spitsbergen population of bowhead whales within a few decades. The summer of 1613 saw seven English ships, seven Dutch ships carrying Basque whalers and twelve ships from the Basque country itself. By 1614 the Dutch fleet of fourteen ships was accompanied by three or four large warships, while thirteen English vessels contended with the Dutch for harbours and convenient locations to establish shore stations for rendering whale oil. This process involved towing the whales to a flat shore where the blubber was flensed and cut into small chunks that were boiled in immense metal cauldrons, set on stone or brick foundations over fires fuelled by the fritters of the last rendering. The resulting oil was cooled in a wooden trough or a whale-boat partially filled with water, and ladled into wooden barrels for transportation.

The air of Spitsbergen was soon permeated with the stench of burning fat, as black columns of greasy smoke rose continuously from most of the accessible harbours throughout the summer months. Dutch and British whalers negotiated, sometimes at gun-point, for prime locations, and in 1617 the Dutch Noordsche Compagnie began to construct a permanent base on tiny Amsterdam Island just off the northwestern coast of Spitsbergen. This was Smeerenburg ("Blubbertown"), a collection of rendering furnaces, storage sheds, workshops and sleeping-huts that gave rise to one of the more attractive myths in Arctic history—that of a small Dutch town among the icefields at the edge of the known world. The legend began to take form shortly after Smeerenburg was abandoned, when the Dutch writer C.G. Zorgdrager reported that

> *All these Cookeries and Warehouses along the flat of Smeerenburg looked like half a small town or village, which therefore was not inaptly called Blubbertown after the industry.... Seeing that the ships, as already stated, brought up double crews, it was very dull, not only on the ships and boats but also ashore. There came up, therefore, as in a camp, some sutlers who sold their wares, such as brandy, tobacco, and the like in their own huts or in the warehouses. Bakers went there also to bake bread. In the morning, when the hot rolls and white bread were drawn from the oven, a horn was blown, so that some enjoyment was then to be had at Smeerenburg.*

By the early twentieth century the Norwegian explorer Fridtjof Nansen had added the raucous flavour of a Klondike town:

> *Here was an entire city with shops and streets.... Some ten thousand people in summer with the noise of packhouses, and the train-oil cookeries, gambling halls, of smithies and workshops, of pedlars and dance halls. Along this flat beach a throng of boats with sailors just coming from the exciting whale-hunt and of women in gay colours on man-hunt.*

This romantic picture has been demolished by closer readings of historical accounts, and by archaeological work carried out during the 1980s by the Dutch scholar Louwrens Hacquebord. The gravel beaches of Amsterdam Island have yielded evidence for the existence of only eight brick furnaces built to support the cauldrons in which blubber was rendered, each associated with one or two brick and timber worksheds and dormitories. Also identified were a blacksmith's shop and the possible remains of a small stone-walled fort. The summer population of Smeerenburg is now estimated to have been only about 200 men, far fewer than in fanciful reconstructions of the settlement, and there is little evidence of either the gaiety or the bourgeois comforts described by earlier writers. Several fortunes in oil were produced by the whalers, flensers, carpenters and coopers of Smeerenburg, but for these men the settlement must have been a place of cold, hard work, evil-smelling smoke and a pervasive fear of the alien environment in which they were exiled. In six burial places on Amsterdam Island the archaeologists found the graves of 101 men who had the misfortune to be overtaken by death while working in this bleak outpost.

The whales that fed the furnaces of Smeerenburg and the other establishments around Spitsbergen's coast were soon hunted out. By the 1640s they no longer frequented the bays and inshore waters where they could be hunted from small boats launched from shore, and whaling ships began to launch their hunts in the open ocean. Blubber was increasingly rendered aboard ship, or packed in the hold and taken to shore-factories in Holland. Smeerenburg was abandoned around 1650, but the offshore whaling fleet continued to expand for another half-century. In 1697, a year that is uniquely documented in historical accounts, 201 whaling ships set out from Dutch and Hanseatic ports, 1,968 whales were taken and 63,883 casks filled with blubber. The whale populations of the Norwegian Sea could not support such a hunt, and whalers were soon turning to the unhunted populations of Davis Strait.

A new element in Spitsbergen's history is mentioned for the first time in the 1697 accounts, which report that several Russian ships visited the islands. Tsar Peter the Great was in Holland that summer, where he was treated to a demonstration of a Spitsbergen whale-hunt, but whaling does not appear to have been the focus of the Russians who began to frequent the area in increasing numbers. These were Pomors, seafaring people from the coasts of the White Sea who, like the Cossacks of the south, had maintained their freedom from serfdom. Long before the eighteenth century the merchants of Arkhangelsk had developed a maritime trade with northern Scandinavia, as well as with the English and Dutch whose ships plied the northern routes to Russia. Furs and ivory were the basis of this trade, and these were the commodities that Pomor hunters sought in the Spitsbergen archipelago, which they called *Grumont*, a name apparently derived from "Greenland."

The English and Dutch at Spitsbergen viewed the sea-faring abilities of the Pomors with derision. Their *lodjas* were described as poor and clumsy ships, their skippers as peasant-sailors whose navigational aids were limited to a compass and a hand-drawn chart. These ships must have been more seaworthy and the sailors more skilled than they appeared to western Europeans, for in most years for a century and a half Pomor crews endured the fifty-day crossing from the White Sea to successfully strike this tiny archipelago in the Arctic ice. The *lodja* was probably a bluff-bowed plank-built ship similar to those used in Hanseatic ports, 15 to 20 metres long with two or three masts and a bank of oars; it probably had little keel, since the ships were pulled ashore for the winter. The western Europeans were impressed by the manners of these strange men with huge beards, fur hats and sheepskin coats. John Laing, surgeon on an English whaler in 1806, reported of Pomor visitors that "during the time they were on board, and particularly while at meat, they behaved with a decorum and gentleness which could hardly be expected from their grotesque appearance."

On landing in Spitsbergen, a Pomor crew constructed a simple house of driftwood logs with a clay stove, and spent the summer hunting reindeer and other animals to support themselves over the coming winter. Although European whalers had occasionally spent winters in Spitsbergen, it was usually by accident and almost always ended tragically; the Pomors were the first to make a practice of wintering over. The season of darkness was mostly passed in tiny outpost cabins, barely two or three metres to a side, heated by a meagre fire and a small oil lamp. From these huts they checked their lines of deadfall-traps set for Arctic fox, and desperately fought off the symptoms of scurvy. Excessive sleep was thought to be a primary symptom or perhaps a cause of the disease, and Russian trappers are reported to have spent the winters tying and untying knots in a rope, desperately trying to limit their sleep to five hours a day in order to avoid the ravages of disease.

From this way of life came the first folklore indigenous to Spitsbergen. Scurvy was pictured in Pomor lore as an old hag, whose eleven beautiful sisters enticed trappers to sleep, or into the dangerous interior of the islands where they fell asleep and froze. Another myth of the Russian trappers was the Spitsbergen Dog, a huge and malevolent creature that lived in the interior with one of the beautiful sisters. He was blamed for bad weather and hunting misfortunes, but if hunters gained his favour by sacrificing animals, the Dog would lead them to reindeer and drive foxes into their traps. The Dog was also said to be fond of drink, and was known to send storms to deliberately sink a ship so that he could swim out and tow the floating casks of spirits to his warm cavern in the interior of the country. This comfortable refuge, furnished with brandy and shared with one of the beautiful sisters, must have haunted the dark thoughts of many Pomor trappers as they endured the endless winter night in a cramped hut filled with cold and hunger and the temptation of sleep. Scurvy, freezing and starvation carried off many entire wintering-parties, but the survivors often seemed to develop a fatal attraction for the country. An old Pomor crew-chief named Starashchin passed more than thirty winters on the island, spending the last fifteen years of his life in permanent residence; he died there in 1826.

The Russian Pomors had Spitsbergen to themselves for much of the eighteenth century, but Norwegian hunters began to appear in the area about 1820. Sailing first out of Hammerfest and then from more southerly ports, commercial hunters loaded their small sloops with reindeer hides, eiderdown and walrus ivory. The increased competition in European markets, as well as the decline in the reindeer and fox populations caused by the intensification of hunting, brought an end to the Pomor ventures. The last recorded wintering by Russians was in 1851–52.

Walrus ivory was the most valuable resource for nineteenth-century Norwegian hunters. Walrus populations had survived the whaling period, when the bowhead whale had diverted hunters' attention from smaller sources of sea mammal oil. The walrus seem to have modified their behaviour, however, spending more of their time in the water or hauled up on drifting pack-ice rather than on the rocks and beaches where they had been so vulnerable to earlier hunters. When isolated groups did come ashore, they rarely survived. Sir James Lamont, in his book about an 1858 sporting trip to Spitsbergen, *Seasons with the Sea Horses, or Sporting Adventures in the Northern Seas*, describes a Norwegian walrus hunt on High Rock Island:

In August, 1852, two small sloops approached the island and discovered a herd of walrus, numbering many thousands, reposing on it. Four boats' crews (16 men) proceeded to attack. The great mass of the walrus lay in a sandy bay with rocks enclosing it on each side and on a little mossy flat above the bay, but to which the bay formed the only convenient access for such unwieldy animals. The boats landed a short distance off, and the 16 men, creeping along the shore, got between the sea and the bay full of walrus, and immediately commenced to stab the animals next them. The walrus, although so active and fierce in the water, is very immovably and helpless on shore, and those in front soon succumbed to the lances. The passage to the sea soon got blocked up with dead and dying walrus. When drenched with blood and exhausted, and their lances from repeated use became blunt and useless, they returned to their vessel, had their dinner, ground their lances, and then returned killing 900 walrus. Unable to carry all the bodies back to Norway in their sloops, the hunters hurried back to Norway with part of their booty, and returned later in the season, but they found the island cut off by ice, and so were obliged to leave hundreds of carcasses.

Lamont's most vivid descriptions relate to the open-water hunting of walrus using harpoons, guns and lances from boats. He reports that "In all my sporting life I never saw any thing to equal the wild excitement of these hunts." The most effective technique was to capture or wound a baby walrus, and to torture it for as long as possible; its "barking grunts" of pain attracted the rest of the herd to its defence, where they could be slaughtered at close quarters. Lamont was very impressed by the actions of a mother walrus, harpooned and towing a boat, while trying to protect the calf which she carried under a flipper by interposing her body between it and the questing lances. "I never in my life witnessed any thing more interesting and more affecting than the wonderful maternal affection displayed by this poor walrus."

Lamont's interest and affection did not prevent him from joining in the sport of walrus hunting, although he reported that "they are undergoing a rapid diminution of numbers" as a result of the toll taken each year by the commercial hunters. Indeed, the hunt soon ended with the disappearance of the walrus, and Norwegians turned to the reindeer herds that still roamed the interior tundra. The Pomors had killed reindeer only for food, but the Norwegians developed a meat-hunt for export to Norway, and within a few decades the Spitsbergen reindeer had almost been exterminated. The eggs of migratory birds became another export commodity, together with the down collected from eider nests. In the summer of 1921 Gordon Seton met a pair of Norwegian hunters who had collected 15,000 eider duck eggs and were daily gathering hundreds more; the eggs were packed in barrels of sawdust for sale in Norway, and the down sold for bedding. Having found this Arctic oasis already stripped of its whales and walrus, bears and reindeer and foxes, these late-comers had to content themselves with the last and smallest species that could be destroyed for profit. Sitting on the shore of Isfjord, Seton wrote of a land that was "fairy-like, ethereal, yet so far as the eye could see, entirely devoid of life.... And everywhere was the silence that broods ceaselessly about the lands that approach the Pole." This was a man-made silence, and one that had required over three unrelenting centuries of effort to attain.

Four years after Seton's visit the Spitsbergen Treaty came into effect, recognizing Norwegian sovereignty over the islands. The archipelago began to be known by the Norwegian name Svalbard, and the treaty ended the "tragedy of the commons" that had encouraged the wanton destruction of its animals. The hunting of bear and walrus was forbidden, and the taking of other species regulated by law. Coal became the new industry, and the easily mined deposits that had occasionally fuelled Dutch rendering-furnaces and Pomor stoves now began to be exploited in commercial quantities. In 1916 the Norwegian company Store Norske opened mines at Longyearbyen in Isfjord, and this town of about 1,500 is today the administrative and transportation centre of the archipelago. In 1931 the Soviet government bought a Dutch mining concession at Barentsburg, 40 kilometres to the west of Longyearbyen, and Barentsburg remains today a Russian town of about 1,000 people, complete with the world's most northerly farm and dairy. The strategic position of the archipelago was probably of greater importance than the profits of coal-mining in maintaining these communities throughout the period of the Cold War, but since 1990 there have been increasing contacts between the Russian and Norwegian communities. Both are surviving despite financial setbacks, and shocks such as the 1996 air crash that killed many wives and children of the Russian miners at Barentsburg.

Tourism is the most visible economic activity in contemporary Svalbard. Brightly-clad travellers now hike the interior valleys, skirt the coast in sea-kayaks, and are landed from cruise-ships to savour this fragment of Arctic Europe. They search for bears, walrus, whales and birds, and consider themselves fortunate to encounter any remnant of what were once vast populations of these creatures.

I visited Svalbard in the late summer of 1986, just 390 years after it had first been sighted by Willem Barents. I was also aboard a Dutch ship, this one named *Plancius* after the great Renaissance geographer. The small Rotterdam pilot-boat had been converted for use as a research vessel by the Arctic Centre at the University of Grøningen, and was also used to carry groups of Dutch tourists on Arctic cruises (this was the type of cruise in which the passengers took turns cooking and washing-up). I was aboard at the invitation of Louwrens Hacquebord, the archaeologist whose work at Smeerenburg had

Walrus bones carpet the beach of an old killing site on a Svalbard coast. (Robert McGhee, Canadian Museum of Civilization)

opened a window on the history of Dutch whaling in Svalbard. We were undertaking a dubious search for evidence that Neolithic hunters might have penetrated this far into the Arctic, as they had in most other circumpolar areas by about 5,000 years ago.

We found no traces of such ancient hunters, but the history of the past four centuries lay all about us. The captain took us to three sets of rendering-furnaces that he had discovered on small islands off the coast of Edgeøya, remarkably well-preserved structures, that had probably been built and used by seventeenth-century English whalers. We frequently came across the weathered remnants of small wooden buildings, sometimes only a scatter of grey boards or poles, marking Pomor wintering sites. Everywhere in the interior were the remains of wooden deadfall traps set centuries ago for Arctic foxes. In Hornsund a dilapidated cabin had been repaired by a Pole and a Czech who were reliving a Pomor adventure. Six Arctic foxes, as tame as cats, scampered around the cabin soliciting handouts; it was unclear what lay in store for them during the coming winter when they had developed valuable winter coats.

The most common and most poignant remnants of the past are the bones of the animals that first attracted human activity to these distant islands. Massive whale bones protrude from the muddy beaches where the animals were once flensed. In every valley, tundra flowers and lichens slowly cloak the fragile bones of reindeer. And then there are the walrus kills, encountered unexpectedly on every coast. For hundreds of metres the surface is carpeted with thick and heavy bones, the massive skulls with missing tusks remaining as impenetrable to decay as they were to seventeenth-century bullets. The drifts of bones are thickest near the beach, where the hunters created a windrow of dead and dying animals to prevent their relatives from escaping to the sea. At a killing-place just north of Kap Lee I wandered inland through a tangle of walrus bones that formed a great triangle, its base along the beach and its point ascending the steep mountainside. Here, 200 metres from the sea, the last remnants of the herds had vainly tried to outclimb the men with iron lances. I clambered among the bones with a dreadful sadness, surrounded by centuries of fear and pain, miserable at the thought of these clumsy sea-beasts hunted on a mountainside.

Some of the animals have returned to Svalbard. Tundra ponds are alive with summer waterfowl, and the nesting-cliffs of seabirds are torrents of noisy life. Reindeer graze the valleys and Arctic foxes patrol the beaches. The dogs belonging to the weather station on Hopen Island had chased 220 polar bears during the winter before my visit. Yet bowhead whales are rarely seen, despite almost a century of protection from hunting; as elsewhere in the Arctic, there is little hope that they will recover their former numbers. The tourists aboard *Plancius* had been promised a sighting of walrus, which began to reappear in the islands during the 1970s, but none had shown themselves during our tour. On our last day we made an overland hike to a cove where the captain had always found a small group hauled up and resting. Our crew was uncharacteristically quiet as we climbed the ridge behind the cove, and carefully peered over the top. The sunny beach below was deserted, and somehow desolate.

10 BAY OF TRAGEDY

HUDSON BAY IS A VAST, FROZEN SEA that plunges like an icy wedge towards the heart of North America. Its waters carry Arctic seals and whales, ice and polar bears, deep into the forests of central Canada. Its southern tip lies less than 700 kilometres north of where I am writing, on the Ottawa River at the edge of urban North America. The geese resting on the autumn river before continuing their southern migration may have left its freezing shores only yesterday. Dropping out of the northern sky towards the first cornfields they have seen all summer, the wavering lines of tired and quiet birds form a tangible link with that distant northern world. The Bay plays a vital role in the Arctic mythology of Canadians, reflecting the essential part that it occupied in history as the focus of the Hudson's Bay Company, the fur trade and two centuries of western exploration.

I grew up trying to imagine the great frozen ocean that lay just beyond the northern forests of Ontario. More intriguing was the mythical region extending 1,000 kilometres from its western coast, the great triangle of tundra, crystal rivers and caribou herds that for three centuries has been known as the Barren Grounds. This was the Arctic that Farley Mowat experienced during the 1940s, and wrote of with such passion and humanity. His books led me to those of

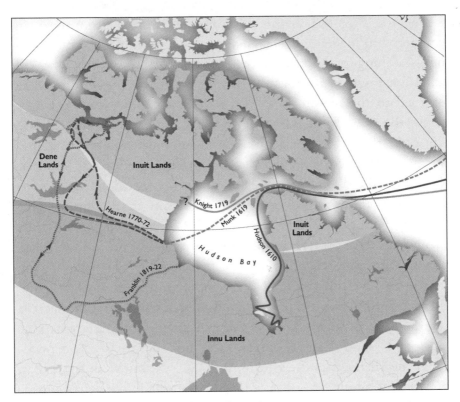

Exploration of Hudson Bay and the Barren Grounds

other barren-ground travellers, particularly to Samuel Hearne's astonishing account of his eighteenth-century trek across the Barrens. Hearne's book takes one on an intimate journey through a world so different from any known today that it might be a particularly vivid fantasy imagined by his contemporary Samuel Coleridge. Going even further back in time, almost two centuries before Hearne and only a generation after Martin Frobisher's adventures on Baffin Island, other accounts of Hudson Bay described some of the most pathetic tragedies in the history of European ventures to the Arctic. When I first caught sight of the Bay during the summer of 1962— an endless panorama of drifting icefields stretching away under the wing of the bushplane that had just picked me out of a summer job in the hot flat forests behind the James Bay coast—I was looking at the theatre in which played some of the best-known dramas in Canadian history.

The tragic pattern began with Henry Hudson, a journeyman explorer who made four remarkable voyages in four successive summers searching for a northern route to Asia. In 1607 and 1608 he had sailed for England in Spitsbergen waters and the Barents Sea; he returned to the region in 1609 for the Dutch, but ice and unrest among the crew diverted him westwards to America and the Hudson River. All three voyages demonstrated his outstanding seamanship and nose for discovery, but faults in Hudson's personality must be blamed for the dissent and threatened mutiny that dogged the ventures. In 1610 he sailed from England again in the tiny barque *Discovery,* with a crew of twenty-two men, intent on penetrating the channel off the northern tip of Labrador that Martin Frobisher had named Mistaken Strait. This he accomplished, surviving the fog and ice and massive tides of what is now known as Hudson Strait, and discovering that this turbulent stretch of water was the gateway to a huge and almost landlocked sea. *Discovery* sailed southwards along the rocky and island-studded eastern coast of Hudson Bay, and penetrated the shallow southern extension that we know as James Bay. With a continuous water-horizon to the west and no means of measuring longitude, Hudson probably thought that he had reached a northern sector of the Pacific Ocean, and was travelling southwards along the western coast of America.

The only surviving account of the voyage from this point on was written by Abacuk Pricket, whose report suffers from the fact that he was not a mariner but a representative of the investors. He describes several weeks of apparently aimless wandering in James Bay and southern Hudson Bay, perhaps reflecting Hudson's puzzled attempts to reach the latitude of California, which Francis Drake had attained from the south three decades before. At the southern end of the bay Hudson was only about 10° to the north of Drake's Nova Albion on the coast of California, and the same distance north of the point at which Hudson had given up his exploration of the Hudson River the previous summer. The hot August days that occur when winds sweep off the immense forests surrounding James Bay must have reminded him of the sweltering summer weather of the Hudson Valley, emphasizing its proximity. Surrounded by forested coasts, and

at a latitude that was about the same as that of London, he may have
had little apprehension of spending the winter. When ice began to
form in early November the crew hauled *Discovery* ashore and prepared for winter, planning to supplement their six months' provisions
with fish and game from the surrounding forests.

But James Bay lies in the heart of the Subarctic, with winter
temperatures averaging between −10° and −25°C; most animals flee
the windy coast for the shelter of the forested interior, as do the
Cree hunters whose land they were visiting. Through the entire
winter Hudson's crew met and traded with only one native, and the
lack of generosity with which he was treated probably explains the
fact that he did not return. The insecurity of their food supply,
together with the constraints of passing a long and cold winter in
confined quarters, must have aggravated the tensions that had
begun to develop between crew and captain. Hudson's rages and
vacillations, his erratic displays of favouritism, together with
rumours that he was hoarding food, contributed to the unrest.
Although the crew killed many ptarmigan throughout the winter,
spring saw them searching the forest for moss and even frogs. The
doctor made a tea from tree buds that cured their scurvy, and their
nets brought in enough small fish to maintain them until the ice
broke up in June, but their attempts to find the native inhabitants
of the area in order to buy meat were unsuccessful. Abacuk Pricket
notes "To speake of all our trouble in this time of Winter (which
was so cold, as it lamed the most of our Company, and my selfe doe
yet feele it) would bee too tedious."

By the time that *Discovery* was released by the ice, and they were
pushing north out of James Bay, the crew was sharing out the last
morsels of ship's biscuit and mouldy cheese. They were stopped by
the icefields that cover much of Hudson Bay during the early summer, and again faced starvation. Hudson had a search made of
personal belongings, which produced a quantity of pilfered and
hoarded food as well as much bad feeling. Pricket's description of the
developing mutiny, and of the actual events as they happened, is the
most detailed, immediate and moving account of such an event that
I have encountered. The extent of its candour is arguable, since

Pricket was a survivor and his journal was written with a legal defence in mind, but the general story is clear.

The strongest men had apparently decided that the sick and weak must be abandoned, together with the captain and his supporters, in order that the others should survive. The plan was to force the victims of the plot into the boat that was being towed by *Discovery*, and to cut them loose. On June 23 the plot was put into effect:

> *In the meane time Henrie Greene, and another went to the Carpenter, and held him with a talke, till the Master came out of his Cabbin (which hee soone did) then came John Thomas and Bennet before him, while Wilson bound his armes behind him. He asked them what they meant? They told him, he should know when he was in the Shallop….Then was the Shallop haled up to the Ship side, and the poore, sicke, and lame men were called upon to get them out of their Cabbins into the Shallop…*
>
> *Now was the Carpenter at libertie, who asked them, if they would bee hanged when they came home: and as for himselfe, hee said, hee would not stay in the Ship unlesse they would force him; they bad him goe then, for they would not stay him….The Carpenter got of them a Peece, and Powder, and Shot, and some Pikes, an Iron Pot, and some meale, and other things. They stood out of the Ice, the Shallop being fast to the Sterne of the Shippe, and so (when they were nigh out, for I cannot say, they were cleane out) they cut her head fast from the Sterne of our Ship, then out with their Top-sayles, righted their helme, and lay under their Fore-sayle till they had ransacked and searched all places in the Ship.*

The picture of the captain and his young son, the faithful carpenter and six helplessly ill men abandoned to a certain death in the icefields of Hudson Bay is one of the most pathetic in Arctic exploration history. No trace of the shallop or its crew has ever been found. We can imagine them working their way northward through the ice along the eastern shore of Hudson Bay, with hopes of retracing their route to the coast of Labrador and finding help among the European fishermen who visited that coast. They would almost certainly have encountered the Inuit who occupied the area, and whom they would have seen as a source of food and possibly salvation. For these Inuit the wooden boat carrying a few ill and helpless men would probably have been too valuable a prize to resist, and the encounter probably ended as did that of Frobisher's five lost sailors, and also that which the mutineers experienced a month later.

In late July, hungry and lost, the surviving crew of *Discovery* were elated to find themselves back at the bird-nesting cliffs that they had named Digges Cape at the northeastern corner of Hudson Bay. The cliffs on Digges Island and adjacent Cape Wolstenholme still support a breeding colony of over half a million thick-billed murres, forming one of the most impressive locales in the region. The mutineers found the place occupied by Inuit, hunting and preserving birds taken from the cliffs with nets on long poles. The Inuit were happy to trade meat and walrus tusks for mirrors, bells, buttons and knives, but during a second visit ashore a trading session turned into an ambush as the unarmed Englishmen—including the chief mutineers—were attacked with knives and arrows. The English fought to get their boat clear of the shore, but the ringleader Henry Greene was killed immediately while three others died in agony from their wounds during the following day. This bloody event served to clear the others of the crime of mutiny, and when the eight surviving men reached England—after an Atlantic crossing in which they subsisted on murre skins and candles—they managed to escape prosecution as well as starvation.

The London merchants who had financed the Hudson venture seem to have intervened in the course of justice, in order to retain the services of men who knew the way into the newly discovered sea to the west of Labrador. The following year Abacuk Pricket and Robert Bylot, the mate who had brought *Discovery* home in 1611, were aboard another venture into Hudson Bay, and suffered through another winter dogged by starvation and scurvy passed at the mouth of the Nelson River on the forested southwestern coast of the bay. Their adventure was repeated in a more drastic form seven years later in the winter of 1619–20, when the Danish skipper Jens Munk was sent by the King of Denmark to exploit this new and promising route to a Northwest Passage.

Munk was an officer in the Danish/Norwegian navy, and is remembered in Scandinavia as their first great Arctic explorer. The son of an impoverished and disgraced aristocratic family, Munk went to sea at the age of twelve and underwent a variety of adventures in places as distant as Brazil. By 1605 he was captaining ships in northern trade and in the Barents Sea explorations for a Northeast Passage. Among his northern exploits were the organizing of the first Danish whaling expedition to Spitsbergen waters and the 1615 capture of a notorious pirate off the northern Russian coast. Munk's combination of experience and well-demonstrated competence made him the obvious choice to command a royal expedition in search of a Danish Northwest Passage.

Munk set sail from Copenhagen in May of 1619 with two ships, *Unicorn* and *Lamprey*, and sixty-four men. He followed the ancient route to Greenland, due west from the southern end of the Faeroe Islands to strike the glaciered east coast of Greenland, then south with the ice to round Cape Farewell. Finding Davis Strait encumbered with ice and fog, he entered the mouth of Frobisher Bay but quickly realized that, despite the contrary opinion of his English pilot, this was not the route to the west. He soon found Hudson Strait, took accurate latitudes of the islands that flanked its entrance to north and south, and made detailed notes on tides, currents and ice conditions for those who would later follow his route. In what must have been a meticulous or very lucky piece of sailing, he deliberately rammed an icefloe in order to repair a heavy section of *Unicorn's* frame that had been displaced by earlier ice pressure, and was able to report that "This was done as perfectly as if twenty carpenters had been engaged in refitting it; all that remained to be done was to adjust the bolts that were bent." Landing on the southern coast of Baffin Island, the crew traded successfully with a party of Inuit, and hunted caribou and hares before continuing to the west. Some time later they were caught helplessly in the ice, and not only prayed but—a first in the annals of exploration—took up a charitable collection for the poor. Released from the ice at the western end of Hudson Strait, Munk set a course to the southwest and sailed for three days and nights on the 1,000-kilometre crossing to the southwestern coast of the bay. He arrived on September 7, too late in the season to do anything but establish winter quarters in hopes of undertaking a full season of exploration the following summer.

Munk selected a wintering location in the mouth of the Churchill River, one of the largest rivers flowing into Hudson Bay, and his log tells of elaborate precautions undertaken to secure the ships against ice. The location lies at treeline, and the surrounding forests—probably more extensive than those found there today—promised fuel and shelter. The latitude of 59° N was about the same as that of Munk's childhood home in the mild climate of southern Norway, and the fierce cold of the subarctic winter that was coming was probably beyond their expectation or imagination. On September 12 there is a brief note in the log that appears ominous in the light of what was to happen:

Early the next morning a large white bear came down to the water's edge, where it started to eat a beluga fish that I had caught the day before. I shot the bear and gave the meat to the crew with orders that it was to be just slightly boiled, then kept in vinegar overnight. I even had two or three pieces of the flesh roasted for the cabin. It was of good taste and quite agreeable.

Undercooked polar bear meat is a dangerous source of trichinosis, a parasitic infection that in mild cases causes fatigue, headaches, diarrhea and muscle pains. If the infection is heavy enough—which depends on the number of parasites infesting the original host—the symptoms can last for months and can even result in death.

The autumn passed in orderly activity. The small *Lamprey* was closed up for the winter, and the crew moved into the larger *Unicorn*, where Munk had two great fireplaces built to provide warmth. Men went ashore every day to collect fuel, hunt, trap foxes for their fur, and pick the same tundra berries with which they were familiar from Norway. Wine was rationed for special occasions, but the men were allowed to drink as much beer as they wished. On November 21 Munk's journal noted that the weather was "as fine as could be expected in Denmark" for the season, and also that "we buried a

sailor who had been ill for a long time." On December 20 in mild
weather the entire crew was ashore, gathering fuel and hunting fresh
meat for the Christmas holidays. Christmas Eve was celebrated with
wine and strong beer, which had to be thawed but of which the men
drank as much as they could hold. A Christmas Day sermon and mass
was held and a Christmas offering was collected, but recorded with
an ominous note: "There was not much money among the men, but
they gave what they had. So many of them gave white fox skins that
the priest had enough of them to line his coat. A life long enough
to wear it was not granted him, however." The crew played games,
and Munk reported that "the crew was in good health and brim-
ming with merriment."

The deaths began in early January. As temperatures plunged to the
low point of the year a violent illness quickly spread through the crew,
beginning with dysentery and ending three weeks later in death.
Munk's journal gradually becomes a daily litany of death. By the middle of February only seven men were capable of collecting wood and
water, and hunting for the odd hare and ptarmigan to make soup for
the invalids. By March the sickness was compounded by scurvy, and
the survivors began to be too weak to bury the dead. On March 30
Munk recorded "My greatest sorrow and misery started at that time,
and soon I was like a wild and lonely bird. I was obliged to prepare
and serve drink to the sick men myself, and to give them anything
else I thought might nourish or comfort them. I was not accustomed
to such duties, however, and had but little knowledge of what should
be done." Opening the medicine chest, Munk found that all of the
medications were labelled in Latin; not only were they inscrutable to
him, but he guessed that the late doctor must have had to depend on
the late priest for translations. In mid-April only four men besides
Munk were capable of sitting up in their berths to hear the Good
Friday sermon that he read. By May 10 when the geese arrived from
the south, only eleven men were still living to taste them.

On June 4, with only four men alive, Munk wrote a note requesting anyone who found them to bury the bodies, and to transmit his
journal to the King "in order that my poor wife and children may
obtain some benefit from my great distress and miserable death.

Herewith, goodnight to all the world; and my soul into the hands of God..." Four days later Munk was the last man alive on the ship, and used his last strength to crawl on deck away from the stench of bodies thawing in the summer warmth. The following morning he was astonished to see two men still alive on the nearby shore, men who had been too weak to return to the ship and had been presumed dead. They helped him ashore, and the three survivors crawled about the forest eating anything green that they could find. With this nourishment they gathered the strength to set a net that produced several trout, to retrieve wine from the abandoned *Lamprey* and to hunt geese.

Scurvy was beaten back and their strength soon returned. By the end of June they had managed to refloat the *Lamprey* and transfer supplies from the death-ship *Unicorn*. When the ice broke on July 16 Munk and his two shipmates set sail and saw the last of their tragic wintering site. For a month they contended with the ice of Hudson Bay, and for another month with Atlantic gales, finally arriving on September 21 in a small harbour on the southern coast of Norway. Munk closes his narrative with the straightforward report that

> *On 25 September I arrived at Bergen, where I went immediately to a physician to obtain advice and remedies. I also arranged for drinks and medicine to be taken to my two men by the skipper who was to replace me on the Lamprey. Then, on 27 September 1620, I wrote to the high authorities in Denmark to report that I was home.*

Safe in Bergen, Munk must have found it difficult to believe that less then four months before he had written that other note, surrounded by dead and dying men, requesting only burial from whoever found the message. What must have been the worst experience of an exciting life was behind him, but cannot have been easily forgotten during the remainder of his distinguished naval service. The following year he was ordered to prepare another expedition to Hudson Bay, but this endeavour collapsed before it sailed. Eight years later Munk was dead of wounds suffered in an engagement undertaken as part of the interminable quarrel over Schleswig-Holstein.

The journal that Munk published as a record of his Northwest Passage adventure tells a grim story, a tale of death on a scale that was rarely surpassed among exploration accounts. Yet the reader is not left with the impression of dismal incompetence and personal ineptitude that pervades the accounts of so many expeditions gone wrong. Munk was no Hugh Willoughby or Henry Hudson. The clothing, equipment and supplies probably could have been better, but may well have been the best available given the knowledge of the time, and there is no indication of food shortage or injury caused by inadequate clothing. Problems were documented for remedy in future ventures: Munk noted that skis would have been useful for winter mobility, medicines should not have been stored in glass bottles that shattered when frozen, and of course they should be labelled in Danish. No blame can be easily assigned for the tragic events that unfolded at the mouth of Churchill River in the winter of 1620, only bad fortune and the medical knowledge of the time that didn't understand the causes or cures of scurvy and the other afflictions that beset the crews with such devastating effect.

After the Munk tragedy the English had Hudson Bay to themselves. By 1670 the Company of Gentlemen Adventurers Trading into Hudson's Bay (later the Hudson's Bay Company) had obtained a royal charter to the territory of Ruperts Land, comprising the land drained by all of the rivers flowing into the bay. As in the Siberia of the time, it was fur that attracted Europeans to the forests of Subarctic Canada, and fur trading posts were soon established at the mouths of rivers draining into James Bay and the forested southwestern coast of Hudson Bay. From these tiny posts indigenous trade routes extended far into the interior of western Canada, and the fur trade soon stretched to the Rocky Mountains and beyond. Ships made annual journeys between England and Hudson Bay, and world commerce extended into the heart of the continent.

At some time during the early part of the seventeenth century a very unusual Arctic expedition may have been undertaken, a venture that seems to have been almost entirely overlooked by history. The single account of the event appears in the writings of Pierre-Esprit Radisson, the French fur trader whose defection to England was crucial to the establishment of the Hudson's Bay Company, and the English intrusion into what had been a French monopoly in the northern North American fur trade, Radisson, who had lived and travelled widely among the native communities of the Great Lakes region, reported that at some time in the past a group of Huron Indians had searched for new land in the north. Supplied with a year's worth of provisions, they set out from their homeland in the farming country of what is now southern Ontario, and travelled northwards through the river systems to James Bay. Here they built canoes large enough to hold thirty men and set out northwards along the coast, occasionally meeting small groups of timid people whose language they did not understand. They left the forests and entered a barren country that was all rocks and hills; travelling further they encountered mountains of ice, and then reached a country which Radisson identified as "the golden arm." Continuing along this coast they found the weather becoming warmer, and were eventually surprised to find themselves once more among familiar peoples in the estuary of the St. Lawrence River.

Radisson was a man who loved a good story, and told some tales of his own travels that are difficult to believe. Before dismissing this story as fantasy, however, we should consider a few pieces of related information. The Hurons were an Iroquoian farming people whose lands commanded the trade routes between the French in the St. Lawrence Valley and the northern Great Lakes forests. Until their destruction about 1650 by a combination of European disease and wars with their Iroquois relatives, the Hurons were consummate merchants, carrying French goods into the interior and exchanging them for beaver skins. Their commercial contacts undoubtedly extended to the rivers flowing into James Bay, and they or Iroquois successors visited the area at least occasionally, as I learned in the summer of 1962 when I travelled with Cree fishermen from the community of Nemiscau to the east of James Bay. One day we were traversing the long portage leading from the Rupert River system to Lac Evans— several kilometres of boggy trail across which we carried canoes, gill-nets, outboard motors, gasoline, tents and everything else required for a summer's fishing. In the late afternoon as we crossed a rocky

ridge, my companions offered to show me a place where their ancestors had been massacred by southern Indians, and where their bones
still lay. I was simply too exhausted from the unfamiliar labour of carrying heavy equipment through the heat and insects of the portage to
be interested in history, a decision I have long regretted. The local oral
history is supported by the account of an early Jesuit missionary, who
reported an Iroquois fort on an island in Lake Nemiscau.

Neither the fort nor the massacre is likely to have been directly
related to the exploration story told by Radisson, but both do confirm a Huron or Iroquois presence in the country around James Bay.
The barren hilly landscapes that the voyagers are said to have encountered as they went north could easily describe the tundra coast of
eastern Hudson Bay, and following that coast their route would have
continued along the southern shore of Hudson Strait. From here they
would have reached the coast of Labrador, a name that derives from
its sixteenth-century discoverer, João Fernandes, who was a *lavrador* or
small land-holder on the Azorean island of Terceira. By Radisson's
time this derivation must have been replaced by the French folk ety-
mology of "le bras d'or," the "golden arm" that has been a mystery to
readers of Radisson's story (which was written and published in
English.) The same process has occurred with regard to Cape Breton
Island's Bras d'Or Lakes. The Labrador coast would have brought the
Huron voyagers into the estuary of the St. Lawrence River, where
they would have encountered the French traders and settlers with
whom they were familiar, and eventually their Huron compatriots.
Despite the flimsy evidence that such an expedition was actually
undertaken by Huron explorers, there is a curious consistency underlying the story. The voyage would have been long and extremely
hazardous, but the Huron traders and warriors of the Great Lakes had
experience with large bodies of water, hostile local populations and
travelling on long journeys through unfamiliar country. This rare
example of a northern exploration by a non-European people has
not, to my knowledge, ever been cited in books dealing with Arctic
history. Perhaps it should not be so easily discarded.

A century after Munk's terrible winter at the mouth of the
Churchill River, the region saw another venture that had an equally

disastrous but more mysterious end. In 1717 a trading post had been established at the location by James Knight, a man in his sixties with long service to the Hudson's Bay Company. Knight had heard rumours from native sources that he interpreted as indicating that both gold and a Northwest Passage existed somewhere along the Hudson Bay coast to the north of the Churchill River. He persuaded the Company to provide him with two ships, the *Albany* and the *Discovery,* and to outfit an expedition in search of wealth and fame. The ships and their crew of about forty men sailed in June of 1719 and were never seen again. The first indication of their fate surfaced almost fifty years later, by which time the Company was regularly sending small ships north from Churchill to hunt whales and to trade with the Inuit of northern Hudson Bay. One such vessel landed on Marble Island, an eerily bleak quartzite outcrop lying 500 kilometres to the north of Churchill and 15 kilometres off the coast, where they discovered the remains of a house, a scatter of anchors and other items too large to have been salvaged by Inuit, and the hulls of two ships sunk in the shallow water of an adjacent harbour. Two years later the crew of the Company ship met Inuit in the area who told them a pathetic story. One autumn, they had encountered about fifty men, *qadlunaat*, who were building the house on Marble Island. When they visited the following spring there were many fewer, and by that autumn only twenty unhealthy men survived. That winter the Inuit lived nearby and supplied the *qadlunaat* with meat and blubber. When they returned after hunting on the mainland in the spring only five were alive, and three of these died soon after eating the raw seal meat and blubber the Inuit gave them. The final two survivors spent their time on the top of a nearby hill, looking to the south for the arrival of help, but eventually one perished and the other died while trying to bury him. The Inuit named this white outcrop Dead Man's Island, and began avoiding it as a place of illness and misfortune.

This sad tale was recorded by Samuel Hearne, who was mate of the ship that visited Marble Island in 1769. A quite different story seems to emerge from an archaeological excavation of the locale that was carried out in the 1990s under the direction of Owen Beattie.

The lower walls of Knight's house, built of stone and turf, were still standing, and the deposits inside the structure and in the surrounding area were much as Knight had left them. The remains of the sunken ships were found to be in excellent shape, and salvage work seems to have been undertaken while they were still afloat, which would rule out shipwreck as a cause of the tragedy. Remnants of a large pile of coal indicate that fuel had been successfully offloaded, so that its lack had not been a problem for the wintering party. Numerous animal bones seemed to show that they had been well supplied with fresh food. The only human bones recovered were a single vertebra found outside the house, and a couple of loose teeth from the interior. John Geiger, a writer who participated in and reported the archaeological work, questions Hearne's account of what he had been told by Inuit, and concludes that most or all of Knight's party did not die on Marble Island. Perhaps they disappeared while attempting an exploration of the Barren Grounds on foot. A vague story had reached the Churchill River post in 1721 that Indians in the far interior had met Europeans from whom they traded iron. It is more likely that the men abandoned their ships, which were locked in the ice of Marble Island's small harbour, and attempted to walk south to the trading post 500 kilometres away. The elderly James Knight might not have expected to survive such a trek, yet there are no signs of his remains on Marble Island. What does remain is the local Inuit custom that visitors to the island must crawl ashore on hands and knees to avoid sickness and misfortune, a tradition that may in some manner relate to an ancient tragedy.

A half century after James Knight wintered on Marble Island, a remarkable explorer set out from the post at the mouth of the Churchill River. The tiny post that Knight established in 1717 had now developed into Fort Prince of Wales, a massive masonry structure built to withstand French contenders for ownership of Hudson Bay and access to its valuable trade in furs. The explorer was Samuel Hearne, who had served during the previous summer as mate of the ship that visited Marble Island, and who had reported the Inuit account of Knight's fate. An ordinary seaman, twenty-four years of age and relatively new to the fur trade, Hearne was not obviously cast

Fort Prince of Wales, the Hudson's Bay Company post at the mouth of the Churchill River,
as it appeared at the time of Samuel Hearne's travels. (From Samuel Hearne, *A Journey
from Prince of Wales's Fort in Hudson's Bay to the Northern Ocean, 1769–1772*, edited by
Richard Glover. Macmillan: Toronto, 1958)

in the mould of the great explorers, and in fact may have been
given his task simply because he was considered expendable by the
Governor of the fort. Yet over the following three years Hearne
traversed the Barren Grounds as far west as Great Slave Lake and as
far north as the Arctic coast. His reports provided detailed information on a vast stretch of country previously unknown to Europeans,
and his description of a shoal-and-ice-filled sea far to the north
destroyed any hope that a practical Northwest Passage would be
found at a conveniently low latitude.

Hearne's achievement is often undervalued or even ignored by
historians of Arctic travels. This may in part be due to the assumption
that Arctic exploration is primarily a nautical pursuit; even the great
British overland expeditions of the nineteenth century were naval
endeavours, the officers and crews lacking only the familiar deck
beneath their feet. Hearne went exploring by himself, attaching himself to native hunting and travelling parties and putting himself at the

mercy of their leaders. In doing so, and in writing of his travels with perception and understanding, he produced a unique and detailed picture of the Barren Grounds and their occupants before they were changed forever by the great smallpox epidemics of the late eighteenth century. He also gave the world a very rare account of the way of life enjoyed by an aboriginal Arctic population before they were significantly influenced by involvement in the global culture that was even then spreading about the world.

Before we follow Hearne onto the Barren Grounds and onward to the tragic event that was the climax of his journey, the stage should be set with the peoples who formed the main characters in his story. Three aboriginal nations were involved in the early fur trade in Hudson Bay. Most prominent were the Cree (whom Hearne called the Southern Indians), a mosaic of closely related groups that occupy the great northern snow-forests stretching from the Atlantic coast of Labrador to the foothills of the Rocky Mountains. Forest hunters of moose and caribou, fishers of lakes and rivers, travellers by toboggan and birchbark canoe, the Cree were among the first to become involved in the European fur trade. Supplied with European goods to trade with more distant peoples, and with guns to ensure their advantage over unarmed groups, they soon spread westward to handle much of the trade from the Canadian Prairies and the forests to the north. Fort Prince of Wales at the mouth of the Churchill River was at the northern limit of Cree territory.

The other nation that traded at Fort Prince of Wales were the Dene, Hearne's Northern Indians. Under various local names given them by the traders (among others, Chipewyan, Slaves, Copper Indians) these were the most easterly of a series of related groups that stretched through Alaska and almost to Bering Strait. The languages and the genetic patterns of the Dene peoples set them apart from all other aboriginal American populations, suggesting that their ancestors may have been among the last to move to the hemisphere from northeastern Asia. The eastern Dene with whom Hearne travelled were peoples of the forest edge and the tundra regions to the north, their lives centred on the migratory patterns of the caribou, which were their main sustenance and the focus of their lives.

Samuel Hearne carved his signature into a rock near Prince of Wales Fort two years before setting out on his explorations of the Barren Grounds. (Fred Bruemmer)

Finally, far to the north lived the Inuit. By Hearne's time a tentative trade had been established with Inuit who lived along the west coast of Hudson Bay to the north of Fort Prince of Wales. The massacres of European crews in earlier times had not been forgotten, and relations between the English and the Inuit continued to be wary, tempered only by the great mutual advantage of a trade in furs furs for ironwork. Relations between Dene and Inuit had not advanced beyond a mutual hostility and mutual contempt, the results of which were to be demonstrated to Hearne with great directness.

Hearne was given his orders in the fall of 1769, shortly after he had returned from his trip to Marble Island. The Governor of the fort had decided to send an expedition to search out the distant source of copper from which the Dene made their knives and weapons, and the rumoured nearby sea. The leader of a small Dene band was provided with supplies and compensation to guide Hearne and four companions—two Englishmen and two Cree—to the copper mines and the northern ocean. They set out in early November, when the country was well frozen and there was still considerable daylight to facilitate travel. However, soon after they left the fort it became apparent that the Dene leader had no intention of fulfilling his part of the agreement, and after two weeks Hearne and his companions abandoned the venture and walked over 200 kilometres back to Fort Prince of Wales, arriving in December. In February of 1770 he set out with another party, but by August it was clear that the guide had no idea where he was going; nor could he protect Hearne when the group was plundered of most of their belongings by a stronger party of Dene. Hearne gave up and returned to the fort, arriving in late November after having travelled over 1,000 kilometres in eight months.

Hearne set out on his final journey after only two weeks of rest, accompanying a Dene leader named Matonabbee who had a reputation for competence and trustworthiness, six hard-working wives to carry his possessions and an extensive knowledge of the distant country to the west and north. On this journey Hearne took no Cree or English companions and for the next two and a half years he was immersed in the experience of a Dene hunting band. For a literate European, it was an unparalleled opportunity to observe and record the life of an aboriginal Arctic people.

Barely two weeks after leaving the fort, Hearne had his first lesson in the insecurity that relentlessly shadowed people who depended on this bleak land for their lives. Hearne and his companions found themselves starving in the rocky country just north of treeline, at a season when the winter sun barely grazes the southern horizon at noon and average temperatures fall between $-20°$ and $-35°C$. He expressed admiration for the fortitude of the Dene, and of the manner with which they treated such appalling situations:

December 27 1770: On the nineteenth, we pursued our course in the North West quarter; and, after leaving the above-mentioned creek, traversed nothing but entire barren ground, with empty bellies, till the twenty-seventh... [when they shot four caribou, and stopped to gorge on the meat]. Indeed for many days before we had not tasted a morsel of any thing, except a pipe of tobacco and a drink of snow water; and as we walked daily from morning till night, and were all heavy laden, our strength began to fail. I must confess that I never spent so dull a

Christmas.... My Indians, however, still kept in good spirits; and as
we were then across all the barren ground, and saw a few fresh tracks
of deer, they began to think that the worst of the road was over for that
winter.... I have more than once seen the Northern Indians, at the end
of three or four days fasting, as merry and jocose on the subject, as if
they had voluntarily imposed it on themselves; and would ask each
other in the plainest terms, and in the merriest mood, if they had any
inclination for an intrigue with a strange woman? I must acknowledge
that examples of this kind were of infinite service to me, as they tended
to keep up my spirits on those occasions with a degree of fortitude
that would have been impossible for me to have done had the Indians
behaved in a contrary manner, and expressed any apprehension of
starving.

For the next four months the party gradually worked its way
westward along the treeline, living from the caribou that at this time
of year stayed close to the protection of the forests. Periods of famine
and plenty alternated, and Hearne noted and understood the need to
travel constantly in such country in order to survive. He also understood the waste of food that was part of this way of life, and remarked
that when the Dene killed a large number of caribou,

... notwithstanding we frequently remained three, four or five days in
a place, to eat up the spoils of our hunting, yet at our departure we
frequently left great quantities of good meat behind us, which we could
neither eat nor carry with us. This conduct is the more excusable among
people whose wandering way of life and contracted ideas make every
thing appear to them the effect of mere chance. The great uncertainty of
their ever visiting this or that part a second time, induces them to think
there is nothing either wrong or improvident in living on the best the
country will afford, as they are passing through it from place to place,
and they seem willing that those who come after them should take
their chance, as they have done.

A few months later he was less philosophical about Dene killing
practices. When he attempted to convince his companions that they

should not kill large numbers of caribou merely for the marrow and tongues, and especially at a time of year when the skins were of no use,

> ... I was always answered, that it was certainly right to kill plenty, and live on the best, when and where it was to be got, for that it would be impossible to do it where everything was scarce: and they insisted on it, that killing plenty of deer and other game in one part of the country, could never make them scarcer in another. Indeed, they were so accustomed to kill every thing that came within their reach, that few of them could pass by a small bird's nest, without slaying the young ones, or destroying the eggs.

Hearne was puzzled by, but sympathetic to some of the social customs by which the Dene lived. He understood that men needed as many wives as they could support, since women undertook all of the work involved in butchering, preserving and preparing meat; sewing and maintaining clothing and tent-coverings; and in the transport of the family's children, possessions and supplies in their endless travels. In turn, women required the support of a dependable hunter and competent planner in order to maintain themselves and their children on the near side of starvation. Women and other possessions were taken from, or lost to, other Dene groups that they occasionally encountered, and on one of these occasions Hearne was shocked that Matonabbee casually stabbed a man whose wife he had taken by force. He noted

> It has ever been the custom among those people for the men to wrestle for any woman to whom they are attached; and, of course, the strongest party always carries off the prize. [Boys practise wrestling constantly, as this]... enables them to protect their property, and particularly their wives, from the hands of those powerful ravishers; some of whom make almost a livelihood by taking what they please from the weaker parties, without making them any return.

Hearne was amazed that such incidents very rarely turned violent, and that all parties usually escaped without physical injury. He couldn't understand Matonabbee's action, given that he was "in every other respect, a man of such universal good sense, and…such great humanity," and blamed it on Matonabbee's having lived for so long among the Cree.

The stabbing occurred in May when many small bands, totalling about 200 people, had gathered at a place where there were birch trees large enough to provide bark covering for canoes. This was obviously an annual gathering, at a time of year when the rivers were breaking up and travel was difficult, and when everyone worked at preparing tent poles and canoes for the summer on the treeless Barren Grounds. By this time the caribou were moving out of the forests, the pregnant cows leading the trek to calving grounds near the Arctic coast. The Dene and the wolves followed the herds northward into a world of tumbling rivers and partly frozen lakes, clouds of insects rising from the soggy tundra, and vast open distances. Hearne found his party much enlarged, but was concerned to learn that many men had joined the group "with no other intent but to murder the Esquimaux," who were known to occupy the northern sea coast to which Hearne was bound. When Hearne protested this plan he was derided as a coward, and knowing that not only his journey but his life depended on the continued goodwill of his hosts, "never afterwards ventured to interfere with their war-plans."

In early June the party began moving northward, crossing lakes on the spring ice and later using their canoes to ferry across rivers. They had now caught up with the caribou herds, and were able to kill as many as they wished. Somewhere to the east of Great Bear Lake they encountered a band of local Dene whom Hearne called Copper Indians, who had never before seen a European. Hearne had little to trade with these strangers, and Matonabbee's large group plundered them of women, furs and even their bows and arrows, although as usual there was no overt violence and Matonabee managed to have most of the weapons returned to people who would die without them. Several of the Copper Indian men joined the war-party as guides to the hills where copper was found, and to the nearby lands of the Inuit where they hoped to join in the attack.

Leaving the women, children, dogs, tents and most other possessions at the Copper Indian camp, where they had prepared a supply
of fresh and dried meat, the war-party moved quickly northward to
the Coppermine River. Through sleet and snow so constant that they
could rarely light a fire for comfort, and then through heat and
clouds of mosquitoes, they trekked for two weeks, sleeping in the
shelter of rocky cliffs when they could find them and in the open
when they could not. Spies sent in advance reported five Inuit tents
at a set of rapids several kilometres above the mouth of the
Coppermine River. The disorderly gang of Dene hunters was transformed into a disciplined party of warriors under Matonabbee's
command; they painted their faces and shields, and lay in ambush
until the Inuit had settled to sleep. They then rushed on the camp
and quickly completed what has become one of the most famous
massacres in Canadian history. Hearne was shocked by the brutality
shown by men whom he had grown to trust and respect:

> *My situation and the terror of my mind at beholding this butchery, can-
> not easily be conceived, much less described; though I summoned up all
> the fortitude I was master of on the occasion, it was with difficulty that I
> could refrain from tears; and I am confident that my features must have
> feelingly expressed how sincerely I was affected at the barbarous scene I
> then witnessed; even at this hour I cannot reflect on the transactions of
> that horrid day without shedding tears.*

Despite his protestations, Hearne managed to describe the action
in close and horrific detail; the account is so vivid and has been so
often reprinted that the massacre at Bloody Falls has epitomized for
many readers the brutality of aboriginal life in the Arctic. Another
view has recently been proposed by literary scholar I.S. Maclaren,
who notes that the crescendo of violent detail occurs only in the
final 1795 published version of Hearne's account, and suggests that it
may owe more to the influence of the "gothic" sensibility of the
period than to what Hearne actually saw. But even shorn of its more
macabre elements the account is chilling, and Hearne's tearful reaction is compatible with what would today be diagnosed as

post-traumatic stress. Five families of Inuit died that night at the place that has come to be known as Bloody Falls, and their bones lay on the surface fifty years later when John Franklin's expedition descended the Coppermine River on their way to the Arctic coast. Two centuries after the event I camped for several weeks at Bloody Falls, and although the area of the massacre had been heavily overgrown with dwarf willows, the remains of ancient Inuit camps were still visible. It is still a marvellous summer fishing spot, where migrating char can be speared or gaffed from the rocks at the foot of the waterfall. It was here that the final victim was killed: an old woman, too blind and deaf to be aware of the nearby commotion, who continued fishing until she was surprised by the warriors. Standing on that particular ledge of rock, deafened by the roar of rushing water, one uneasily contemplates the human cruelty that permeated this region two centuries ago.

The remainder of Hearne's journey was long but anticlimactic. Bloody Falls is only about a dozen kilometres above the mouth of the Coppermine River, and the day after the massacre he walked downriver far enough to see seals on the ice of the island-filled sea that stretched to the northern horizon. On the return southward he was guided to the hills where copper was found, but it was clear that this thin and scattered deposit would be of no commercial value. Hearne was told the local Copper Indian tradition that explained the lack of copper: it stated that the mines had once been extremely rich when they were first discovered by a woman, and for several years she had led men to the locality where they gathered metal for all their needs. When some of the men took liberties with her she vowed revenge and, being a great shaman, sat on top of the hill that contained the mine and gradually sank into the earth together with all but a few remaining scraps of copper. Two centuries later, on my way back from working on the Coppermine River, I sat in the bar of Yellowknife's Gold Range Hotel. It was at that time a friendly and sometimes scary room, stripped of everything but small arborite tables scattered across a linoleum floor under fluorescent strip-lights. I was sharing a table with a Yellowknife Dene man who, when he heard that I had been on the Coppermine, told me the identical story

Bloody Falls, a few kilometres above the mouth of the Coppermine River, seen from downstream. Hearne's party travelled from the hills in the far distance, and the massacre occurred on the sloping ground just to the right of the waterfall. (Robert McGhee, Canadian Museum of Civilization)

about the disappearing mine and added the detail (perhaps suppressed by Hearne) that the woman's revenge was in repayment for being raped by her brother. I bought a glass of beer for the man who had not only confirmed the strength of oral tradition, but whose great-grandfather five or six times removed had probably been with Hearne at Bloody Falls on that still July night in 1771. I felt again what I had on the rock ledge at Bloody Falls: that in local history the past is often closer than we assume.

Hearne spent another winter with Matonabbee's party, travelled through the forested country around the eastern arm of Great Slave Lake, and eventually returned to Fort Prince of Wales in the early summer of 1772. Having walked over 2,000 kilometres and lived dangerously close to starvation for four years, Hearne concluded his report by stating

Though my discoveries are not likely to prove of any material advantage to the Nation at large, or indeed to the Hudson's Bay Company, yet I have the pleasure to think that I have fully complied with the orders of my Masters, and that it has put a final end to all disputes concerning a North West passage through Hudson's Bay.

Despite his failure to find either a profitable copper mine or a Northwest Passage, Hearne advanced in the Company's service and ten years later was Governor of Fort Prince of Wales when a French fleet under the Comte de la Pérouse arrived. With the wisdom gained from participating in naval engagements during his youth, Hearne realized that the fort was indefensible and surrendered it without a fight. He and his men were repatriated to England, together with the manuscript of his travel journal that La Pérouse read and returned to him on the condition that it be published. Finding the fort destroyed and his English patrons gone, Matonabbee hanged himself; his wives and children starved the following winter. The pattern of tragedy continued to haunt the early Europeans and those who dealt with them around the rim of Hudson Bay.

11 FROZEN GLORY

SAMUEL HEARNE'S REPORT on his astonishing journey across the Barren Grounds put an end to lingering hopes that a commercially viable Northwest Passage existed across the top of the American continent. The rumoured fabulous deposits of yellow metal in the distant northwest had turned out not to be gold, but scattered useless chunks of native copper. As in Siberia, any wealth that was to be obtained in the northern regions of North America would still come from the fur-bearing animals of the great snow-forests that spanned the continent. Fur continued to drive the exploration of the Subarctic. In 1789, the same year that Russian merchants finally established a trade-fair in the lands of the fiercely independent Chukchi, the fur trader Alexander Mackenzie reached the coast of the western Canadian Arctic a few hundred kilometres to the east of Chukotka. Four years later, when he penetrated as far as Russian territory on the northern Pacific coast of Canada, the fur trade had spanned the world.

With no immediate prospects of wealth to be gained in the tundra landscapes and frozen oceans to the north of treeline, the commercially driven Arctic explorations of past centuries came to an end. New purposes and new players began to transform the field of

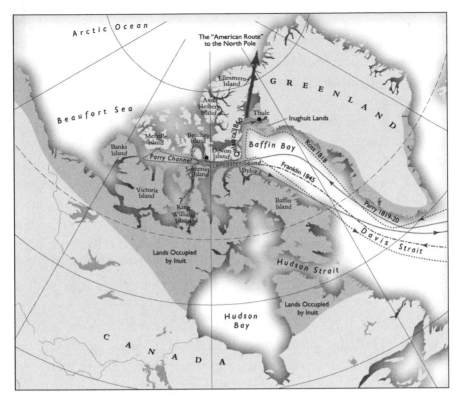

Exploration of the High Arctic

Arctic exploration. Patriotic fervour, the desire to expand human knowledge and most prominently the quest for personal advancement and personal celebrity formed a potent mix of motives that attracted romantics, misfits and megalomaniacs to the Arctic regions throughout the nineteenth and twentieth centuries.

The manner in which explorations were reported also underwent a profound change. The publication of mariners' logs and the journals of explorers had been first developed, by men like Richard Hakluyt and Samuel Purchas, as a means of disseminating information that would stimulate commercially valuable national endeavours. This tradition continued into the eighteenth century, but Hearne's 1795 book on his Barren Grounds journey, together with Mungo Park's immensely popular 1799 book on his African adventures, were early examples of a new form of travel narrative. These books were not published as sources of commercial information, but for the enjoyment and enlightenment of the literate public. They emphasized the romantic description of remote places and strange peoples, together with the reaction of the explorer to these phenomena and to the privations and dangers that he underwent to encounter them.

The nineteenth century saw a massive production of such literature, developing as both a byproduct and a stimulus of the period's European colonialist expansion. When they were done well—as in the case of both Hearne and Mungo Park—these books became lasting classics. Many others, despite their initial success, have not appealed to the sensibilities of a later time. When compared to the appallingly misspelled but crystalline prose of Elizabethan mariners, and the spare, thoughtful writing of explorers such as Hearne, the huge books produced by Victorian travellers often feel like wading through deep, wet snow. Of the narratives produced by the mid-nineteenth-century Franklin search expeditions, literary scholar Maurice Hodgson writes that "There was much posturing in Victorian letters, and sometimes the arctic experience brought out the worst." Such writing survived Victoria, reaching a zenith in the publishing industry that accompanied the polar travels of the early twentieth century.

The first major thrust of nineteenth-century Arctic exploration was concentrated on the Northwest Passage, a potential route that had been transformed from a possible source of wealth to a patriotic challenge for English mariners. With too many officers and too many ships left over from the Napoleonic wars, the British Navy sought means to justify its existence and the continued expenditure of taxes on maritime enterprise. The Arctic began to be touted as a training-ground for the seamen and officers that would be required for future wars. Individual officers saw exploration as a means of coming to the notice of their superiors, and of accelerating careers that had been stalled through lack of military action. In his elegantly argued book *I May Be Some Time: Ice and the English Imagination*, Francis Spufford has documented the mixture of class-bound imperial assumption and juvenile homophilic fantasy that characterized British Arctic exploration of the period. This was the environment that gave rise to a century of increasingly costly ventures, organized under the pretext of augmenting geographical knowledge and enhancing sovereignty over portions of the polar world.

The startling difference between earlier explorations and the military-based expeditions of the nineteenth century is very clearly apparent when we compare Samuel Hearne's travels across the Barren Grounds in the 1770s with those undertaken by John Franklin exactly fifty years later. When Hearne travelled from Prince of Wales Fort on Hudson Bay to the Coppermine River and back, he did so in the sole company of Dene families to whom the Barren Grounds were home. He carried no supplies and, aside from a sextant, no equipment that wasn't used by the Dene, and he lived entirely from the country. The journey took most of two years to complete, but the party underwent no more hardship than was normal for Dene life, and Hearne came out of the experience with an unparalleled understanding of both the country and the people who inhabited it.

Franklin set out from York Factory on the southwestern coast of Hudson Bay in the summer of 1819, accompanied by three other naval officers—John Richardson, George Back and Robert Hood—and one seaman, John Hepburn. They were furnished with letters of credit to the Hudson's Bay and Northwest companies, and these fur-trading concerns organized their travel along the canoe and snowshoe routes that had been developed since Hearne's time. A state of commercial warfare existed between the two major companies that would be amalgamated in 1821, and as a result the posts in the far interior were more poorly supplied than usual. The traders had little surplus either to provide for the needs of the expedition or to honour the credits given by Franklin to the native hunters on whom he was to depend for meat. Nevertheless, the party was assisted westward to Cumberland House on the Saskatchewan River, where they wintered, and the following spring northward to Fort Chipewyan on Lake Athabaska and onward to the tiny outpost of Fort Providence on the north shore of Great Slave Lake. In the summer of 1820 the Englishmen struck out towards the treeline, transported by French-Canadian and Iroquois voyageurs paddling the huge birchbark canoes that were the engines of the Canadian fur trade. In the forested country between Great Slave and Great Bear lakes they built a wintering camp that they named Fort Enterprise: a large log building for the officers, a cabin for the voyageurs and the seaman Hepburn, and a storehouse. This was the land of the Dene, and the site was not far from where a war party of these people had joined Samuel Hearne and his companions half a century before (Hearne had known them as the Copper Indians). Now Franklin depended on their descendants to hunt for the expedition and to stockpile food for the coming trip to the Arctic coast.

The following spring they set out, guided by the Copper Indians as far as the site of the Hearne massacre at Bloody Falls, a few kilometres above the mouth of the Coppermine River. Here the Dene guides turned back, leaving the Englishmen, thirteen Canadian voyageurs and two Inuit interpreters to their Arctic explorations. With provisions for only ten days, the party of twenty men in two large canoes set out eastward along the coast of Coronation Gulf, hoping to find a route to Hudson Bay. Four weeks later on August 17, after a storm that frightened the voyageurs (and probably the officers as well), Franklin decided that it was time to turn back. He had explored about 300 kilometres of the mainland coast and the shore continued to trend northeastward without any sign of an outlet toward Hudson Bay. Instead of retracing the coastal route that had been fraught with winds and ice, he decided to strike directly overland for the wintering camp at Fort Enterprise, approximately 500 kilometres away across the Barren Grounds. This decision was to produce a journey of terrible hardship, descending gradually and with a seeming inevitability into horror. Franklin's description of the retreat to Fort Enterprise compares, in its sense of accumulating tragedy, with the account of Barents's wintering on Novaya Zemlya in the earlier search for a Northeast Passage. Like the Barents account, Franklin's had a profound effect on the image of the Arctic in the minds of European readers.

On August 26 the party left the coast and began to ascend a rocky and tumultuous river that they named after Robert Hood, carrying or dragging their large and cumbersome canoes. These were abandoned after a few days and were converted into two smaller canoes, each light enough to be carried by a single man, that would be needed for crossing rivers. On August 31 they set out on foot for Point Lake on the Coppermine River, about 250 kilometres away. Each of the men carried equipment and supplies weighing ninety pounds (forty kilograms), which was the standard burden for voyageurs in the fur trade. The four officers carried "their own things as their strength would permit." The party travelled at a rate of about fifteen kilometres per day.

Franklin's account of the retreat across the Barren Grounds portrays a microcosm of the class-based social system that permeated British Arctic exploration of the period, and depicts the consequences when the system came under unbearable stress. To Franklin the Canadian voyageurs were incredibly strong and hardy brutes of burden, at times cheerful and amusing but at others depressed and malignant, untruthful, fearful of the unknown, incapable of rational planning and constantly requiring a firm directing hand. Their role was to paddle, carry, provide for themselves as much as possible and survive. Natives were capricious and untrustworthy, incapable even of taking direction from someone of superior culture; their role was to be cajoled and enticed into providing the food that only they could harvest from the country. English officers were to be as unburdened as possible in order to be free to navigate, direct and engage in some sporting hunting for the pot when the opportunity presented itself. (British naval officers had evolved far beyond Martin Frobisher, who had toiled with a miner's pick in the freezing hard-rock trenches of Baffin Island 250 years before.) The single seaman, John Hepburn, had the task of providing and caring for the officers.

Snow fell on September 1 and on September 5 a snowstorm stopped their progress for two days; the larger of the two remaining canoes was abandoned after being repeatedly dropped—intentionally, the officers suspected, so that its bearer could be released from his burden. The country they traversed had almost no animals save for a few ptarmigan and hares. They ate the last of their pemmican, felt hunger for the first time, and as starvation-food they began to gather a vaguely edible lichen that the voyageurs called *tripe de roche*. On

September 10 a muskox was shot, providing the first good meal in six days. Two days later they were travelling through wet snow 60 cm deep and had eaten the last of their meat, when they found their course obstructed by the vast expanse of Contwoyto Lake. Here they discovered that their fish-nets had been discarded by the men delegated to carry them, and the floats had been burnt as fuel, so the resources of this fish-rich lake were unattainable. For the following week they tramped slowly through barren rocky landscapes empty of animals, the temperature hovering around the freezing point with wind and snow or rain. Now their rate of travel had dropped to about eight kilometres a day. Richardson abandoned the scientific collections of plants and mineral specimens that he had accumulated throughout the voyage. The men began to talk of throwing away their heavy packs and walking out to the forest by themselves, but apparently were prevented from doing so by the fact that only the officers knew how to find Fort Enterprise.

On September 22 they arrived on the banks of the Coppermine, the last large river that had to be crossed in order to reach the forest and the safety of their winter quarters. Here they learned that their one remaining canoe had been abandoned the day before, having been dropped and broken so many times that the men claimed it was useless. They had now been walking for three weeks during which they were never dry, and their shoes and bedding were usually frozen. All were weakened by the meagre diet of lichen supplemented by occasional small portions of hare or ptarmigan, and some of the party were suffering from chronic and debilitating intestinal problems. For almost two weeks they tried unsuccessfully to cross the river, while the weather turned colder and starvation advanced. They had begun to eat their old moccasins, and to scavenge dead caribou left by wolves, breaking the bones for rancid marrow and even charring the bones so that they could be eaten. It became apparent that those capable of hunting had been eating small game rather than bringing it to camp, and that some of the men had begun to steal the officers' portions of food. The only punishment available was the threat to withdraw future wages, and this had little effect on starving voyageurs. Franklin noted that "Our people... through despondency, had become careless and disobedient, and had ceased to dread punishment, or hope for reward."

On October 4 they finally crossed the river in a makeshift canoe constructed by one of the voyageurs, and reached the first outliers of forest that lay on the other side. Some of the men were incapable of pushing onward through the deepening snow, and began to be left behind. George Back and three voyageurs, stronger than the rest, were sent ahead to reach Fort Enterprise and return with food and help. Hood was unable to go further, and Richardson and Hepburn stopped to care for him and await assistance. Three of the voyageurs were left along the trail the next day, planning to return to Hood's camp. Franklin and four others continued, reaching Fort Enterprise in mid-October to find it cold and abandoned, with none of the long-anticipated stores of food. A number of reasons had conspired against the provision of the expected meat at Fort Enterprise: a poor hunting season; the needs of the Dene to under-take a period of mourning for three drowned hunters; the fact that the trader at Fort Providence had refused to honour the letters of credit that Franklin had given the Dene as payment for obtaining meat; Franklin's disrespectful treatment of Akaitcho, the local Dene leader; and the Dene conviction that the Franklin party was so incompetent that they would perish on the Arctic coast and never be seen again.

A note from Back was found in the empty fort, stating that he had gone in search of the Copper Indians, and intended to try to reach the nearest fur-trade post at Fort Providence but doubted that he had the strength to travel that far. With temperatures now well below zero the freezing men began to dismantle the buildings to feed their fire, and to dig for food among the frozen caribou skins and bones dis-carded during the previous winter.

On October 29 Richardson and Hepburn arrived alone, report-ing that Hood was dead. The next evening Richardson informed Franklin that they had shot Michel Terohaute, an Iroquois voyageur whom they suspected of murdering Hood and of killing and eat-ing others of the voyageurs who had been left behind. Richardson and Hepburn reported that Michel had been acting strangely and had made vague threats to the Englishmen, who became fearful for their lives and decided to forestall him through execution. Franklin was horrified at Richardson's story, and also saddened to learn that of the several men left behind on the trail only the two Englishmen had survived.

The newcomers brought one ptarmigan, which was the first fresh food Franklin and his companions had tasted in a month, but it was not enough to save the lives of the remaining two voyageurs at Fort Enterprise, who died two days later. Assistance arrived the following week in the form of the Copper Indians who had been found by George Back, and who had immediately set out to rescue the explorers. Within ten days they had nursed the survivors to a semblance of health, sufficient to set out with the Dene for the trading post at Fort Resolution on the south shore of Great Slave Lake. From here they made their way home the following summer: the surviving voyageurs to Montréal, the surviving Inuk interpreter to Hudson Bay and the four naval men for England. Behind them they left eleven men dead or missing: one British officer, an Inuk interpreter who had disappeared on the Coppermine River, and nine voyageurs—the class that had borne the weight of the expedition and suffered most of the deaths.

John (later Sir John) Richardson's explanation of the episode at Hood's death-camp has never been entirely convincing or unquestioned. Once the rumour of cannibalism has been raised, it is difficult to confine. In these days, when we know from numerous credible accounts how survivors typically behave in circumstances of Arctic starvation, it is difficult not to suspect that more than one member of the Franklin party had tasted human flesh. An intriguing insight into the reliability of Franklin's extensive published account, and that of the journals kept by the other three officers, is provided by Ferdinand Wentzel. Wentzel was the fur trader who had accompanied the expedition as far as Bloody Falls, and was expected to organize provisions for the returning party at Fort Enterprise; he met the survivors on their return to Fort Chipewyan, and later reported in a letter to his superior that the Franklin expedition had not only resulted in the loss of eleven lives, but that

The archaeological site of Ekven on the northeastern coast of Siberia participated in much of the early development of Inuit culture. The whale bones in the foreground are the remains of an Inuit house; it lies at the top of a mound of artifacts and animal bones that accumulated over several centuries beginning about 2,000 years ago. Cape Dezhneva, the eastern tip of Siberia, extends into the sea at the left, while Bering Strait and the coast of Alaska are hidden in fog. (Robert McGhee, Canadian Museum of Civilization)

Ivory carving of a fantastic sea beast, recovered from a grave in the cemetery on the hill-top behind Ekven. Complex carvings of this sort are characteristic of the Old Bering Sea culture in the centuries around 2,000 years ago, evidence of a society that had the wealth, leisure and iron tools required to develop such a sophisticatred art. (Bryan and Cherry Alexander).

This glaciated island on the edge of the Barents Sea is typical of the inhospitable coasts encountered by early explorers of the Northeast Passage. (Robert McGhee, Canadian Museum of Civilization)

The tents of archaeologists stand amid the slag and charcoal of assays carried out on Qallunaat Island over 400 years ago; one of the mine-trenches cuts across the middle-ground of the photograph. (Robert McGhee, Canadian Museum of Civilization)

Painting by John White representing a skirmish between the Inuit of Baffin Island and a boat crew from Frobisher's 1577 expedition. This encounter occurred on the south coast of Frobisher Bay, but the author has transposed it so as to include recognizable scenery in the vicinity of Qallunaat Island. (British Museum)

A colony of basking walrus on an isolated Arctic island are alert for danger from the photographer's boat. Spitsbergen was once home to hundreds of such groups, but their natural defences proved to be of little use against hunters attracted by the value of their ivory tusks and tough hides. (Frank Todd)

Painting of a Dutch whaling station in the Arctic ("Een Nederlandse traankokerij in de Noordelijke Ijszee") by Cornelis de Man, ca. 1639. Such idealized depictions of industrial whaling probably helped develop the legend of Smeerenburg as a small Dutch town transported to the distant Arctic. (Rijksmuseum Amsterdam)

"Crossing the barren lands," a painting by George Back recording the retreat from the Arctic coast towards treeline during the late summer of 1821. (From C. Stuart Houston, ed., *Arctic Artist: the Journal and Paintings of George Back, Midshipman with Franklin. 1819–1822*. Montréal: McGill-Queen's University Press, 1994.)

The edge of the Barren Grounds in late summer; the first sight of dwarf spruce trees was a vast relief to the exhausted men who had walked from the Arctic coast. (Fred Bruemmer).

Knight's Harbour on Marble Island, the bizarre white island off the northwestern coast of Hudson Bay, where the last traces of James Knight's expedition were found. (Fred Bruemmer)

This 1895 painting by Thomas Smith, entitled "They Forged the Last Link with their Lives," depicts the tragic end of John Franklin's 1845 expedition as imagined from the accounts of the remains discovered by searchers during the late 1850s. (National Maritime Museum, Greenwich)

Tourist cruise-ships now make regular calls at Qaanaaq in far northwestern Greenland, the goal of Qitdlarsuaq's exploration and the region from which Peary, Cook and other polar explorers set out northwards across the ice. (Hans Jensen)

The MI-8 helicopter is a vital communications link in most Siberian communities. This machine is met by villagers at Numto in Khanty Mansiysk, western Siberia. (Bryan and Cherry Alexander)

Saami homes at Kautokeino, Norway, during the polar night. (Bryan and Cherry Alexander)

Pond Inlet, on northern Baffin Island, is typical of the Inuit communities of Nunavut in Arctic Canada. (Fred Bruemmer)

*... the surviving officers have left in the country impressions not alto-
gether very creditable to themselves amongst both the trading class of
people and the native inhabitants. It is to be presumed, as they themselves will be the publishers of the journals which will appear, that they
will be cautious in not exposing their own errors and want of conduct.
In fact one of the officers was candid enough to confess to me that there
were circumstances which must not be known; however it is said that
'stones sometimes speak.'*

Knowing that Wentzel had kept a journal, the officers who had earlier pressed him to visit them in England now "requested me in a
particular manner to remain a year or two more in the country, I
presume with a view to let the storm in some measure subside, or
what is as likely, to take advantage of my not being in the way for
examination..." It has also been noted that the 1821–22 post journal for Fort Resolution, where the survivors passed the remainder
of the winter, has long been missing from the Hudson Bay Company
archives, and a suspicion of cover-up has been hinted.

The spectre of cannibalism was to haunt Franklin's reputation.
When reports of the remains of his final lost expedition began to surface in the 1850s the writing classes of England, including a figure as
notable as Charles Dickens, were horrified and refused to believe that
Europeans, much less British officers, might be capable of such a savage practice. After the first expedition Richardson wrote of Michel
Terohaute that throughout the trip he had been of good and respect-
ful conduct, but that "His principles ... unsupported by a belief in the
divine truth of Christianity, were unable to withstand the pressure of
severe distress." To what extent the repugnance to eating human flesh
is based in religious principles is not clear, nor is it an indisputable
truth that such principles are proof against the degree of distress
experienced by many of the men who followed John Franklin into
the Arctic regions.

Those who have written about Franklin's Arctic Land Expedition
have variously judged it as something between an unfortunate success and a tragic failure. Historian Anne Savours has recently chided
those who blame the tragedy on Franklin's inability to adapt to the

ways of travel and survival used by the Dene and the voyageurs. Would any of these critics, transported to his time, she asks, have done any better? Perhaps not. But surely there were Englishmen of the time who would have carefully read Samuel Hearne's account, and learned from it that survival on the Barren Grounds demanded an intimate knowledge of the seasonal whereabouts of caribou and muskoxen. More importantly, there must have been Englishmen who doubted that the class-based social organization that had grown out of the agricultural land-holding practices of rural England, and had been adapted to its use by the British Navy, could be usefully trans-ferred to the forests and barren tundras of a new world. These are surely the men with whom Franklin, and the other officers who left the frozen bodies of their followers scattered across the Arctic, should be compared in order to judge their fitness as Arctic explorers.

Franklin's overland journey was an unusual naval expedition, a sidelight to the ship-borne British naval exploration parties that pen-etrated the gulfs and channels of the North American Arctic Archipelago. One of the most successful of these was also one of the first. Edward Parry, who sailed with the ships *Hecla* and *Griper* the same year that Franklin left for his overland voyage, chose Lancaster Sound as a likely High Arctic route leading westward from Baffin Bay. With a combination of skilful navigation and good fortune in encountering a season with little sea ice, he sailed west past 113° lon-gitude, and was only stopped from completing a transit of the continent by the masses of polar pack-ice jamming McClure Strait leading to the Beaufort Sea. Parry entered a small protected bay on the south coast of Melville Island where his ships were safely frozen in for the winter, a season that was passed with freshly baked bread and freshly brewed beer, games and theatrical projects, a school for the illiterate sailors, and great care to avoid both depression and scurvy. In the spring he assembled a hand-drawn cart and explored the gravel plains of Melville Island as far as its ice-locked northern coast. The ships were then sawn out of the ice to open water and sailed to England in the summer of 1820. Parry had accomplished a voyage which all hands survived with little hardship, and had attained a position further west than would be achieved by any other ship for several decades. Few Arctic voyagers over the following century could boast such a record of effectiveness, achievement and ease.

Decade by decade, ships continued to cross the Atlantic, contend with the early summer ice of Hudson Strait or Davis Strait, and eventually nose into the maze of channels extending westward into the Arctic Archipelago. Some relied purely on sail while others carried immensely heavy and grossly ineffective steam engines. Most expected to spend at least one winter in the ice, and the supplies provided form a record of nineteenth-century advances in the preservation and canning industries. The skills of ice navigation were only gradually accumulated, and to the end no captain learned how to select a wintering harbour that might not trap a ship when the ice remained frozen during the following summer. An increasing amount of exploration was accomplished by sled-parties that left the ice-locked ships as daylight strengthened in late winter, traversing sea ice and snow-covered land on journeys that gradually traced thousands of kilometres of Arctic coasts.

British naval sledges were immensely heavy vehicles constructed of iron and oak, built to carry up to 1,000 kilograms of equipment and supplies. They were usually hauled by teams of seven men and each was directed by an officer, some of whom put their backs to the traces as well as carrying their own burden of gun, telescope and compass. The men were British seamen and marines, who could be more effectively disciplined than the voyageurs that Franklin had used, but like the voyageurs they were the members of the expeditions who suffered and died in far greater proportion than did their officers. Clothed in impractical costumes of wool, canvas and leather, fed on pemmican and salt pork and biscuit, badly preserved vegetables and a daily tot of rum and lime juice, they faced conditions as terrible as those encountered by any military personnel in world history. In winter sledging the temperatures rarely rose above −20°C and the iron sled-runners grated on snow crystals as if they were sand. As temperatures rose the sleds bogged down in deep snow, and by early summer the men waded through ponds of icy meltwater for hours every day. Men died of hypothermia, gangrene following frozen feet, scurvy and simple exhaustion, and the survivors usually

"Sledging over Hummocky Ice," painted by an officer engaged in the search for Franklin's final expedition. (From S. Gurney Cresswell, *Dedicated...to her Most Gracious Majesty...a Series of Eight Sketches in Colour.* London; Day and Son, 1854.)

returned to England in broken health. That so much was accomplished by such a technique of Arctic travel is a tribute to the endurance and courage of these seamen. The sledge-haulers were both the victims and the true heroes of British Arctic exploration.

The routes of the sled expeditions were marked by caches of supplies, and by rock cairns built as markers or to indicate the location of messages sealed in tin canisters. Many of these humble structures survive today as memorials to the efforts of a past age. Most of the carefully constructed cylindrical cairns have been looted of their messages, some of which are preserved in archives while others illegally grace the private collections of pilots, prospectors and northern administrators. The caches are now heaps of rock and gravel scattered by bears, foxes and curious humans, surrounded by a litter of weathered oak barrel staves, scraps of canvas and rusted tins. During my first summer in the High Arctic I was rounding Cape Majendie in Pioneer Channel, walking on the narrow ice-foot and surrounded by an environment empty of the signs of human activity, when my eye was caught by a flash of red from the rocky shoreline. Going ashore I found a dozen large rectangular tin containers, painted scarlet and bearing printed labels in a blocky stencilled font reading *BACON* and *MEAT*. Some were torn open, while others had large dual punctures made by the canine teeth of a bear. I was mystified by the find, probably because my mind was elsewhere—on the prehistoric archaeological sites for which I was searching, but mostly on the bears that might emerge at any moment from the pack-ice grinding past the point, to find me alone and far from my camp without a gun. Only later it dawned on me that this was a cache left by some almost forgotten sledging party from one of the ships engaged in the 1850s search for the lost Franklin expedition.

The largest scatters of archaeological remains relating to British Arctic exploration are on the shores adjacent to harbours where ships were frozen in for the winter, and at the few locations where ships were abandoned or supplies offloaded in preparation for abandonment. Most of these locations are in regions that were occupied by Inuit, and they served as magnets for people to whom metal goods, wood and the other detritus of exploration were extremely valuable commodities. However, the most impressive such location is at Beechey Island, a tiny island of rocks and gravel at the confluence of Wellington Channel and Lancaster Sound. This region lies to the north of the lands occupied by nineteenth-century Inuit, and the remains of exploration parties were remarkably preserved through the century and a half since the island served as a central focus for the Franklin search expeditions of the 1850s.

The archaeology of Beechey Island reflects most aspects of the bizarre episode that brought an end to the nineteenth-century search for a Northwest Passage. Many books tell the story in great detail, but the pattern of events can be summarized in a few paragraphs. The episode began in 1845 when the British Navy sent its largest and best-equipped exploration party into Arctic North America. The venture was under the command of the same John Franklin who twenty-five years before had travelled overland to Coronation Gulf, and later had led a more successful boat expedition down the Mackenzie River and along the North Alaskan coast. The ships *Erebus* and *Terror* were large ice-strengthened sailing ships with newly fitted auxiliary steam engines. They carried 129 men and supplies for three years, but were almost universally expected to reach the Pacific in two sailing seasons. When 1846 and then 1847 passed with no word that the ships had arrived in Petropavlovsk or San Francisco, concern mounted and the Admiralty dispatched search and relief expeditions from the Atlantic via Lancaster Sound, from the Pacific by Bering Strait, and overland down the Mackenzie River and along the central Arctic coast. Over the next ten years approximately forty further expeditions were mounted, both publicly and privately, but the lost ships were never found.

What the searchers did find were the first wintering quarters, where the ships had been frozen in during the winter of 1845–46 adjacent to Beechey Island. After a decade of searching they finally recovered a message from a cairn on the northern coast of King William Island, almost 500 kilometres to the south, reporting that the ships had been caught in heavy ice in Victoria Strait during September of 1846; Sir John Franklin had died in 1847; and by April 1848 the mariners had given up hope of being released and were planning to abandon the ships and travel south by sledge to the Back River, probably in hopes of reaching a post of the Hudson's Bay Company. Inuit encountered in this sparsely inhabited area reported that the two ships had been sunk in the ice, and that the men had starved to death in trying to escape the region. A trail of skeletons, equipment and personal belongings led southward to the Arctic coast, where all traces of the expedition vanished. Although the massive searches of the 1850s failed to find the lost seamen, their ships or evidence that would provide a satisfactory explanation for the disaster, they did accomplish another goal: that of mapping the interconnected channels that comprise the Northwest Passage, as well as most of the other coasts of the Arctic Archipelago.

Beechey Island gradually became an informal memorial to the lost expedition that had spent its last "happy" winter here, and to the

The ruins of Northumberland House, built by searchers for the lost Franklin expedition on the shore of Beechey Island. The spar in the background probably belongs to the yacht *Mary*, which was left for possible survivors. (Patricia Sutherland)

many search parties that visited the area over the following years. When the site was first located in 1850 the signs of human use included a rock cairn surmounting the island but containing no message, the frozen graves of three seamen who had died during the winter of 1845–46, a pile of 600 empty food tins filled with gravel, and a litter of rope, canvas and other debris. Most of this material was collected by the searchers, who also added the grave of a seaman who died on one of the search expeditions; a wooden storage building named Northumberland House; a small beached ship, the *Mary*, left as support for possible survivors; and a marble monument in honour of the dead. Today Northumberland House is a ruin surrounded by weathered fragments of wood; the yacht *Mary* has disappeared, but a spar that is thought to be her mast leans at an angle from the frozen gravel in which it has been set; the graves have been exhumed and reburied, their invaluable oak markers replaced by replicas. Souvenir hunters have picked the beaches clean of most remnants of nineteenth-century debris, and the site has been polluted by cairns and memorials in dubious taste, commemorating events as trivial as a visit by the Prince of Wales. The site remains sacred to a contingent of amateur historians that still pursue the "Franklin Mystery," each

summer scouring the seabed and beaches of the central Arctic for the
Erebus and *Terror*, the grave of Sir John Franklin or other evidence
that might explain the event that brought an end to the British search
for a Northwest Passage.

Aside from Franklin's ships, surprisingly few exploration vessels
were lost during the half-century search among the Arctic islands for
a Northwest Passage. Yet the fact that most survived is curiously at
odds with the almost constant threats of destruction that are closely
and repetitiously described in published accounts of the voyages. The
ice encountered by these ships must have been rarely as mountainous and hard-driven as described, and was clearly less threatening
than the architecturally implausible forms shown in the drawings and
paintings that illustrate the accounts of nineteenth-century explorers.
These crystalline fortresses, looming under skies that glow with fantastic auroras or blaze with multiple suns linked by arches and rings
of light, heighten the sense of an other-world of unbearable cold and
danger. The British were not alone in creating this make-believe
Arctic world. It was an international literary effort, some of the most
able contributors being the Americans Elisha Kent Kane and Charles
Francis Hall. The descriptions rose to new heights in the subsequent
phase of Arctic exploration, which focused on the attainment of
the North Pole. In this endeavour, Americans dominated, while
Norwegians, Italians and other nationalities played prominent roles.

The beginning of the North Pole adventure coincided with the
end of the Franklin search, and was fed by knowledge and experience gained in the search efforts. It can best be understood as a
redirection of the patriotic and career-building goals of those whose
immediate predecessors had sought a Northwest Passage. The chronicle began with ships working northwards out of Baffin Bay in order
to investigate the possibility that Franklin had taken a route to the
north of Lancaster Sound. The 1852 English voyage commanded by
Edward Inglefield penetrated to north of latitude 78° in Smith
Sound, where Baffin Bay narrows between the converging coasts of
Greenland and Ellesmere Island, and although finding no trace of
Franklin they did report navigable water in the far north. The following year the American Elisha Kent Kane, who was supposedly

looking for Franklin but who also had polar ambitions, pushed his ship *Advance* further north, where it was frozen in and eventually abandoned. Kane's sledge parties reached a new northern record at almost 80° latitude, but the voyage ended in scurvy, starvation and a terrifying escape by boat and foot down the Greenland coast. Kane's party survived only because they made the acquaintance and accepted the help of the local Inuit of northwestern Greenland, the small and isolated band of Inughuit who were the descendants of the people whom John Ross in 1818 called "Arctic Highlanders."

Kane's wildly romantic account of his travels, and of these strange and primitive people, fired the imagination of a wide audience of American readers. Among them was the newspaperman Charles Francis Hall, and two much younger men: Frederick Cook and Robert Peary. Hall would almost immediately set out on his own Arctic adventure, which would end with his death in 1871 on the same coast where Kane had wintered, and a retreat by the remainder of the crew that surpassed even Kane's ordeal in suffering and improbable survival. Cook and Peary would later become the foremost protagonists in a race to the North Pole that was played out during the first decade of the twentieth century, following the route initiated by Kane and relying heavily on the knowledge and support of the Inughuit. Polar exploration by way of the "American route"—the channel between Greenland and Ellesmere Island—was also a major element in shaping the history of these local people, as was another episode that is rarely mentioned in the history of polar exploration, the expedition led by the Inuit shaman Qitlaq, or as he came to be known to his descendants, Qitdlarssuaq, "the great Qitlaq."

This event is well documented in the Inuit oral traditions of northwestern Greenland, and is illuminated by occasional brief mentions in the logs of ships involved in the 1850s search for the Franklin expedition. The leading figure in the story was originally from southeastern Baffin Island, a man who had a history of violence and murder. It may have been as a result of a killing (an account of which was recorded by the anthropologist Franz Boas many years later) that he moved to Pond Inlet in northern Baffin Island. Here he became involved in another deadly fight after which vengeance was expected.

Fearing for his life, he gathered about fifty members of his kin and set out northwards at some time in the mid-1850s. On their way they pillaged a cache of supplies left by a Franklin search vessel, the *North Star*, on a small island near the group's point of departure from the northern coast of Baffin Island. The cache contained twenty-six casks of rum (roughly 1,200 litres) that seem to have been consumed by the Inuit party shortly before they made the dangerous crossing of Lancaster Sound and began their northward trek, placing a wide channel of unstable ice between Qitdlarssuaq and possible pursuers.

The group lived for about five years on Devon Island, a large and partially glaciated island that was otherwise unoccupied, but may have served in the past as a place of refuge for Inuit from Baffin Island. Then, in 1859, Qitdlarssuaq persuaded his followers to set off on a journey of exploration to visit a people he had seen in a vision while on a shamanic voyage, a nation that lived in an unknown land far to the north and dressed entirely in sealskin clothing. The distant people sound very much like the sealskin-clad Greenlandic Inuit, whom Qitdlarssuaq had probably heard of from Edward Inglefield; the British explorer and his Greenlandic interpreter had visited the Devon Island camp in 1853, hoping for information on the missing Franklin expedition. After two years of travel the Inuit group broke into two parties, one of which attempted to return to Baffin Island but apparently died of starvation. The others continued northward, and the following year joined the Inughuit in far northwestern Greenland. In three years the Baffin Island families had successfully travelled more than 1,000 kilometres through uninhabited, unknown and difficult country. Their descendants were prominent members of most of the polar expeditions that set out from northern Greenland between 1860 and 1910. The skill, knowledge and endurance gained by their parents and grandparents during the legendary trek from Baffin Island may have been among the qualities called upon by these Inuit when they became the first humans to travel across the polar pack to the vicinity of the North Pole.

The polar quest added little to human knowledge, but considerably embellished the fantastical vision of the Arctic painted in the minds of southerners. Most of the men who participated in the endeavour wrote tirelessly of the efforts that they undertook, the hazards and hardships that they suffered, and the implacable enmity of cold, darkness and the polar ice. Their narratives are litanies of frostbite and exhaustion, constant thirst, hunger and danger, marauding bears and rough ice that suddenly cracks to reveal lethal pools of black smoking water. The pressure-ridges—massive heaps of ice-blocks as hard as concrete and as difficult to cross as an earthquake-tumbled city—multiply endlessly across every route that is not blocked by impassable leads of open water. Depression, boredom and despair constantly attack the travellers' minds, and their physical condition deteriorates as the effort depletes body fat and then muscle. Surprisingly, most survived to write of the adventure.

After the standard recital of hardships endured, most polar narratives focused on the unique strategies that they employed to— using a metaphor that was common at the time—"climb the ladder of latitudes." The first rung of the ladder was attained in 1871 by Charles Francis Hall, who pushed his ship *Polaris* to a position north of 82°, where the channel between Greenland and Ellesmere Island opens into the polar ocean. The *Polaris* was halted at this point by heavy ice, and the voyage was called off after Hall's murder at the hands of the expedition's doctor. Five years later an English expedition commanded by George Nares wintered the ship *Alert* in the same region, and the following spring Clements Markham set out across the polar pack, using the punishing technique of man-hauled sledges that the English had inherited from the Franklin search period. Markham's sled crew attained 83° latitude before retreating in a confusion of scurvy, snowblindness and breaking ice. This northing was bettered by six kilometres in 1882 by an American military party that travelled by dog-sled from a wintering ship commanded by Adolphus Greely; this expedition ended in starvation and cannibalism.

A new route was attempted by the Norwegian Fridtjof Nansen, who deliberately froze his specially constructed vessel *Fram* into the polar pack to the north of Russia's New Siberian Islands, hoping to drift over the pole before the ship was released into the North Atlantic. When it became apparent that the captive ship would not reach the pole, Nansen and one companion set out over the ice in the spring of 1895, using light dog-hauled sleds. They surpassed 86° before retreating to Franz Josef Land, where they were extremely fortunate to be rescued; the *Fram* itself was released by the ice to the north of Spitsbergen, and returned safely to Norway. In 1897 the Swede Salomon Andrée attempted to drift across the pole in a balloon launched from Spitsbergen, but was forced down on the ice when persistent fog froze to the balloon and weighed it down. The party made their way to an unoccupied island off Spitsbergen where they perished; their remains, together with the journals and photographs describing the terrible episode, were not found until three decades later. Another record northing was attained in 1900 by the Italian naval officer Umberto Cagni, travelling by dogsled from the Duke d'Abruzzi's expedition ship, which wintered in the ice at Franz Josef Land; the Italian expedition was inspired by Nansen's success, and Cagni bettered Nansen's record by thirty kilometres.

The stage was now set for the race to the pole by the "American route," pioneered by Kane and Hall and Greely, which by the turn of the twentieth century was firmly in the grip of Robert E. Peary. Peary was an engineer, an American naval officer with no maritime experience and a man driven by a consuming need for recognition and fame. Writing to his mother in 1880, Peary had stated "I don't want to live and die without accomplishing anything or without being known beyond a narrow circle of friends. I would like to acquire a name which would be an open sesame to circles of culture and refinement anywhere, a name which would make my mother proud, and which would make me feel that I was the peer of anyone I might meet." Through a series of near-disastrous expeditions in Greenland, Peary obsessively accumulated the admiration and financial support of the men whose counterparts a half century later would be known as the American "military-industrial complex." He tirelessly depicted the quest for the Pole as a patriotic endeavour that could be attained only through the right combination of military-like organization and that peculiarly American characteristic known as "dash." The Inughuit sled-drivers and their seamstress wives, on whom Peary depended for transport and survival, were portrayed as cogs in the great machine that he was assembling to roll inexorably to the Pole. They in turn gave Peary a name that translates as "the one who is feared."

The first attempt was made in 1906, and attained a latitude of 87° before being turned back by rough ice and a shortage of food. In 1909 a larger and improved version of the machine was set in motion from Peary's specially built ship *Roosevelt*, which had steamed to the northern coast of Ellesmere Island. The machine consisted of six southerners, eighteen Inuit sled-drivers and 133 dogs pulling several tons of supplies, and was organized so that the sleds would turn back in sequence after they had carried the freight that would support a party making the final dash to the Pole. The last of the support-sleds turned south at latitude 87°46′ leaving Peary, his personal servant Mathew Henson and four Inuit drivers to make the final 220-kilometre sprint. Peary later claimed to have covered that distance in five days, and then to have returned over 800 kilometres to Ellesmere Island in an additional sixteen days. This speed was a surprising record for travelling over polar pack-ice, especially for Peary who had lost his toes to frostbite during an earlier expedition, and who according to Inuit testimony was forced to ride as a passenger for most of the way. This unusual speed was the first element of Peary's story that was questioned when he later made public his claim to have reached the Pole on April 6 1909. This claim was not made immediately, but only after Peary had returned to Greenland and found that another explorer—his fellow American and sometime companion Frederick A. Cook—had declared that he had attained the goal a year previously.

Cook's story told of a much smaller and simpler expedition, financed as far as the Inughuit homeland by an American sportsman. Here Cook, who had previously served as doctor to one of Peary's Greenland expeditions and had also travelled to Antarctica, hired Inuit families and their dog-teams. The group hunted its way across the interior of Ellesmere Island so that the supplies that they carried could be used on the polar ice. From the northern tip of Axel Heiberg Island Cook set out with two Inuit companions and two sleds pulled by twenty-six dogs. The 900 kilometres to the Pole was covered in five weeks, and the group barely survived the southward retreat when food ran out and leads of open water barred their progress. The ice drifted westward and away from the caches of food that they had left ashore for their return, and they eventually landed among the barren islands to the west of Axel Heiberg, where they made their way southward through the melting ice to Devon Island. Having exhausted their ammunition and abandoned their dogs and sleds, they lived by Inuit ingenuity and passed the winter in a refurbished winter house that had been built centuries before by prehistoric Inuit on the coast of Jones Sound. The following spring they walked back to Greenland where the Inuit were welcomed with joy by families who had assumed them to be dead. Cook's report that he had reached the Pole the previous year reached the world only a few weeks before Peary returned with his more recent claim. Peary was appalled by the news, and immediately set out on a life-long campaign to discredit Cook and everything that he had claimed to accomplish. A vicious propaganda campaign financed by the National Geographic Society and Peary's other wealthy backers successfully destroyed Cook's claim as well as his life. Yet no amount of money and influence was able to bury the questions surrounding Peary's own claim to have reached the Pole in the spring of 1909.

In his recent exhaustive book *Cook and Peary: The Polar Controversy Resolved,* Robert M. Bryce concludes that neither explorer is likely to have attained the goal that he claimed. I am not convinced that Bryce is right. It is clear that Peary could not have attained the Pole, and that he was a liar as well as a megalomaniac. Yet when one strips away the false testimony and the biased judgements that emanated from Peary and his supporters, there is no reason to believe that Cook could not have accomplished what he claimed. It is true that Cook was no navigator and may not have been able to precisely identify the location of the Pole, but he certainly could have found his way to the general vicinity. His small and lightly supplied expedition may have seemed impossible in the early twentieth century, but similar trips have since been achieved by others with fewer resources. In 1996 the Canadian-Russian team of Hans Webber and Mikhael Malakhov travelled from Ellesmere Island to the Pole and back, entirely unsupported and hauling all of their equipment and supplies on light toboggans; similar expeditions, some of them successful, are now mounted on almost an annual basis.

Frederick Cook remains the most perceptive as well as the most vilified of polar explorers. In the posthumously published book *Return from the Pole*, he wrote of his discovery that "the greatest mystery, the greatest unknown, is not that beyond the frontiers of knowledge but that unknown capacity in the spirit within the inner man of self.... Therein is the greatest field for exploration. To have suffered the tortures and to have become resigned to the aspects of death as we did—to learn this is experience which no gold can buy. The shadow of death had given new horizons, new frontiers to life." Cook had made a discovery that no amount of humiliation could take from him, and one that only the most fortunate and observant of Arctic explorers ever learned.

12 THE PEOPLE'S LAND

THE NATIVE OCCUPANTS OF THE ARCTIC had little interest in the endeavours of the explorers to which history has accorded such great importance. The quest for sea-passages to the north of Asia and America, the search for Franklin and the race to the Pole, brought relatively few southerners into the Arctic latitudes and very few of those wished to stay longer than was necessary. Most of these visitors found no quick fortunes in the Arctic, had little interest in its people or the lands that they called home, and left the country as quickly as possible to return home and write their books portraying the cold white hell they had experienced. To the northerners who came into contact with such expeditions, the visiting strangers were no more than useful sources of valuable goods: metal tools, wood, exotic ornaments, and eventually tobacco, tea and other staples of the global market. In the latter half of the nineteenth century this situation began to change, as some Arctic peoples saw their countries flooded with hordes of southerners seeking wealth as well as sport and adventure.

These invasions occurred at different times and for a variety of reasons. The eternal quest for whale-oil that had begun in the Bay of Biscay during medieval times moved to Labrador in the sixteenth century, to the Barents Sea in the seventeenth and to Greenland in the eighteenth. As whale stocks were hunted out the whalers discovered unhunted herds in the channels of the Canadian Arctic, and during the latter half of the nineteenth century British and American whalers were the dominant economic force along the coasts of Baffin Island and Hudson Bay. At the same time American whalers began working the waters of the Bering Sea, penetrating Bering Strait by 1870 and hunting along the ice-edges of the Chukchi and Beaufort Seas. The whaling efforts along the coasts of Chukotka and Alaska were an off-shoot of the Pacific and South Seas whaling described so vividly by Herman Melville, and they brought the native occupants of the area into contact with a cast of characters as diverse as that of *Moby-Dick*.

Siberian and Alaskan Eskimos, like the Inuit of the Eastern Arctic, flocked to the shore-stations established by the whalers. Here they found a lucrative market for baleen—the sheets of flexible horn-like material taken from the mouths of whales, known in the nineteenth century as whalebone, and a valuable commodity used in items as var-ied as corsets and carriage-springs. Soon the whalers began to winter in the Arctic, and natives were hired or contracted as hunters to sup-ply the wintering whale-fleets with reindeer and caribou meat, and with skins for winter clothing. Women worked as seamstresses, and men were hired as harpooners or boatmen in the dangerous trade of killing whales from small boats. The availability of such skilled and experienced hunters allowed the captains to inexpensively crew their ships with green hands suited only for manual labour. Some native groups were provided with wooden whaleboats and equipment, and worked as contract-whalers. Food was everywhere abundant, from the caribou and other animals hunted with the rifles introduced by the whalemen, to the carcasses of whales stripped of the commercial products of blubber and whalebone, to the bears and foxes attracted by the bounty of dead whales washed up on every shore. Native communities glowed with a new-found affluence of whaleboats, guns and ammunition, metal pots and pans, knives and stoves, kerosene lamps, sewing-machines and portable phonographs. Tastes and fashions changed abruptly as northerners developed a liking for manufactured clothing, tea and tobacco, home-baked bread and biscuits made from wheat flour, and the spirits so generously served by the whalers at their *hula*-dances. To many Arctic natives the whaling era is looked on nostalgically as a time when their ancestors were valued partners in a lucrative global industry.

Gold and rumours of gold brought the next surge of southerners into the Arctic. The gold fever that had spread from Australia to California in the mid-nineteenth century gradually seeped northward up the mountain valleys of western North America, to flare again in the Klondike strike of the 1890s. Thousands of hopeful miners and hangers-on trekked up the Pacific coast and across the mountains to the Yukon, or travelled to the Bering Sea and up the Yukon River by steamer. Others bought their passage on whaling ships and were put ashore in Eskimo villages, hoping to find their way to the goldfields of the interior. Although the Yukon was the overwhelming focus of the Arctic gold rush, its backwash flowed up many rivers of Alaska and Chukotka and into most native communities throughout the area. The miners brought with them a vision of the Arctic world that combined hatred for what they saw as a blasted and desolate land, fascination with the stark magnificence of absolute cold and silence, consuming greed for the land's hidden wealth, and contempt for the natives who continued to follow what the miners assumed to be an ancient way of life. It was a vision that was easily imposed on the region, and one that outsiders found compelling when it was presented in the writings of Jack London and Robert W. Service. The miner's vision is well expressed in Service's "The Call of the Wild":

> *Have you known the Great White Silence, not a snow-gemmed twig*
>   *aquiver?*
> *(Eternal truths that shame our soothing lies.)*
> *Have you broken trail on snowshoes? mushed your huskies up the river,*
> *Dared the unknown, led the way, and clutched the prize?*
> *Have you marked the map's void spaces, mingled with the mongrel*
>   *races,*
> *Felt the savage strength of brute in every thew?*
> *And though grim as hell the worst is, can you round it off with curses?*
> *Then hearken to the Wild—it's wanting you.*

Few miners attained their dreams of wealth, and most retreated to their homelands after a few years of profitless adventure spent cursing the cold, the natives and the pitiless land. Others stayed in the north with continuing hopes of fortune, and were joined by refugees from the whaling industry. The discovery that petroleum could be refined to produce lamp oil, and spring steel substituted for baleen, put a quick end to the whaling that had in any case almost destroyed the last bowhead stocks of the Arctic seas. As their profits declined, many whalers had begun to supplement their income by trading for other Arctic products: ivory, the skins of bears and muskox and increasingly, as a fashion developed among the furriers of Europe and America, the white winter furs of Arctic fox. The new fashion more than eased the economic depression that was caused in many Arctic regions by the decline of whaling. Trappers in areas with large fox populations became wealthy, their incomes often surpassing those of most North Americans and Europeans. More than a few Eskimos became the owners and skippers of schooners that they used to transport large family groups between their winter trapping areas and the trading posts that opened along many Arctic coasts.

The late nineteenth and early twentieth centuries brought a similar process to the Arctic regions of northern Eurasia. Prospectors drifted northward down the great Siberian rivers, finding a scatter of gold on gravel bars lining the valleys of Yakutia and Chukotka. Norwegian hunters harried the coasts of eastern Greenland, Spitsbergen and the Barents Sea, and encroached on the Russian Pomors' hunting grounds along the coasts of the White and Kara seas. Vast quantities of walrus ivory, sealskins, fox furs, birds' eggs and eiderdown were packed aboard their sloops and exported to the markets of Europe. Some of the indigenous peoples of the area—Saami of northern Norway and the Kola Peninsula, Nenets of the Kara Sea coast—found a profitable role in these hunts, much as did the Eskimos on either side of Bering Strait. In other regions the pressure from European competition was so heavy that native communities were forced to withdraw from their traditional coastal pursuits and join their reindeer-herding relatives on the interior tundras.

Contact with southerners, and knowledge of the lands and inhabitants of warmer climates, was nothing new to Arctic peoples. The Saami had lived in close contact with the Nordic tribes of Scandinavia since at least medieval times, and had long developed mutually beneficial relations based on trade and cooperation in other economic endeavours. The varied peoples of Arctic Siberia had for centuries traded and paid taxes in furs to Russian merchants and government agents. The peoples around Bering Strait had developed an intercontinental trade centred on iron goods that probably arrived from merchants of the Pacific coast to the south. The Inuit of Arctic Canada and Greenland had known the medieval Norse, and had sustained continued contact with European explorers and traders wherever the opportunity presented. This early and enduring participation in the economic systems of the world to the south had long provided northerners with tools and materials that eased their burdens and enriched their lives, while demanding little in the way of change or accommodation.

Yet this process of trade carried the seeds for a more significant alteration in the lives of northern peoples, a change that began to be apparent when Christian missionaries followed in the wake of the merchants. The Christianization of the north succeeded with little of the overt hostility that often characterized the process in more southerly latitudes. From the pagan Norse of the Viking Age to nineteenth-century Inuit, northern peoples who believed in many gods and spiritual beings generally accommodated the new faith that accompanied wealth arriving from the south. For most converts, Christianity was simply a way of gaining access to material goods in exchange for lip service to a strange foreign belief; most aspects of shamanic conviction continued to be the basis of their view of the world. Only gradually and over several generations did people's convictions alter towards those of orthodox Christianity, and even then they retained significant and important elements of older beliefs. Just how far underground aboriginal belief systems have been driven has continued to be an abiding concern of missionaries. It is also a vital question for those whose hopes attach to the survival and revival of native ideologies, and the emergence of new Arctic nations based on old and tested worldviews.

Christianity had an important, though unwelcome, partner in its mission to the Arctic peoples, a collaborator that was extremely effective in promoting acceptance of the new belief. Epidemic diseases, long established among peoples of the temperate world, found new and fertile fields in isolated northern populations that had no resistance to the viruses and other micro-organisms carried by southerners. Missionaries generally arrived in an area when the degree of contact with outsiders was extensive enough to spread disease, and the link between the two arrivals was noted by aboriginal people throughout the world. It was quickly noticed that the shamanic doctors could not cure the illnesses the outsiders had brought.

Norse sagas report epidemics in medieval Greenland following the arrival of ships from Europe, and the pattern continued into the second half of the twentieth century when the arrival of the summer supply-ship in isolated Arctic communities was followed inevitably by a round of colds and sometimes by more serious illness. Epidemics of smallpox and other diseases depopulated much of the Icelandic countryside after the arrival of English fishing fleets in the early fifteenth century, and Arctic peoples of non-European ancestry were even more vulnerable than the Icelanders. Samuel Hearne reported that 90 percent of the Dene population among whom he had travelled across the Barren Grounds in the 1770s were killed by smallpox epidemics a decade later. Smallpox swept through the Inuit communities of Greenland a few years after the arrival in 1721 of the first missionary, Hans Egede. In Chukotka and Alaska, death rates of between 50 percent and 80 percent were reported from many areas during the gold-rush period of the late nineteenth century, and the Inuit of the Mackenzie Delta were almost eradicated by the measles epidemic of 1902. The Spanish influenza pandemic of 1919 killed one-third of Labrador Inuit, and had similar effects in many other Arctic regions.

The effects of epidemics were amplified by the close and crowded conditions under which Arctic peoples lived, the cold and hunger that quickly follow the illness of a provider, and the absence of traditional knowledge related to the treatment of these diseases. The traditional social order of small communities quickly disintegrated in the face of massive depopulation. Confidence in shamans and traditional healers, and the system of beliefs on which their actions were based, collapsed when they failed to deal with frightening new plagues. The acceptance of new religions and new systems of authority by Arctic peoples was helped immeasurably by the epidemics that accompanied increased contact with the southern world.

Despite the dread of disease, the peoples of the north were willing partners in the processes of change that affected their lives through contact with southerners. Arctic peoples freely and enthusiastically exchanged the products of their hunt in return for guns, metal cooking pots, wooden boats and luxuries that made their lives easier and more enjoyable. They learned that some southerners had strange and strongly held beliefs about the creation of the world and the purpose of human life, and were willing to accept that they might be right. But while Arctic peoples became enmeshed in the commercial systems of the larger world, and while they learned about theologies and worldviews imported from distant lands, they continued to see themselves as masters of their own countries and of their own ways of life. No question of sovereignty arose in relation to the lands that they inhabited, nor would such questions have been of any relevance to their assumptions of autonomy and independence in their own domain. For the Siberian peoples who had long paid *yasak* to the Russian Tsars and other rulers before them, and for the Saami who had long been taxed by Nordic kings, these tributes were simply extortion and carried no implication that distant sovereigns owned their lands. Their lands, like those of the Indians and Inuit of Arctic North America, were not property that could be owned, but simply a part of their identities.

The middle decades of the twentieth century brought a series of shocks to this assumption. During the 1930s the new Soviet state extended its power northwards across the breadth of Eurasia. Marx's theory of history, as interpreted by Stalin, demanded that the "primitive communism" of northern peoples be transformed into a more advanced level of social and economic organization. The peoples of the tundra—Saami of the Kola Peninsula, Nenets of the Yamal, Evenki and Ngansan and Dolgans of the Taimyr and Sakha, Chukchi and

Eskimos of Chukotka—were to become members of collectives, a new rural proletariat that owed allegiance to the Soviet state. Cadres of government workers, some backed by soldiers, fanned out across the taiga and tundra, confiscating and reallocating reindeer herds and whaling boats, and establishing local councils that were to wield the authority previously held by the heads of family groups. They built permanent settlements, and boarding-schools where education in the new ways could be provided to the children of people whose way of life had always centred on following the reindeer or travelling from one hunting-ground to the next. Both Christian and shamanic beliefs were discouraged as relics of the primitive past, and as unsuitable to citizens of the new Soviet society that was being constructed.

The discovery that outsiders could exert authority over an entire people, and could do so on the basis of the claim that they owned your land and that you were merely a minor citizen of an immense state, came as a shock to most Arctic peoples. A more intense shock was visited on those whose lands lay directly in the path of industry. Industrialization and the development of northern resources were also priorities of the Soviet state, and the creation of a huge pool of forced labour during the 1930s provided the means for rapidly carrying out grandiose plans. In 1920 the Kola Peninsula had a population of only 20,000 people, most of whom lived near the port of Murmansk, but by 1939 this had grown to 300,000 and major mining and smelting enterprises had been established amidst the reindeer-pastures of the native Saami. The discovery of nickel and other metals at the base of the Taimyr Peninsula led to the 1935 founding of Norilsk. Thousands of prison-labourers were imported to the tundra, surrounding regions were appropriated for hydroelectric reservoirs and transportation routes, and pastures began to be despoiled as plumes of toxic vapours trailed downwind from the smelters. The extensive coal deposits of the Vorkuta Basin began to be exploited in the 1930s, depriving the native Nenets of their traditional pastures and peopling an entire region with convict labourers.

The invasion of northern North America occurred more slowly and on a smaller scale. Placer gold-mining remained the mainstay of southern immigrants in Alaska and the Yukon, but a variety of larger industrial enterprises moved northwards down the Mackenzie River: the 1934 gold strike at Yellowknife on Great Slave Lake grew to a town of 1,000 people within five years; oil drilling began at Norman Wells on the Mackenzie River in 1937 and a refinery was built two years later; the world's richest known deposit of uranium was discovered in 1930 on the shores of Great Bear Lake, and the resulting mine became an important and heavily secured source of vital material during the early development of atomic weapons.

North American governments did not stop to wonder whether they had the right to grant the resources of Arctic regions to those who wished to exploit them for industrial purposes, but their only other expression of sovereignty was through policing. The gold rushes of the 1890s saw the importation of American law to Alaska and of Canadian law to the Yukon, but police were primarily concerned with enforcing this law among the southern miners and others who had flooded into the north. It was a surprise to native people to learn that the law also applied to them, as became apparent in a few famous cases. The killing in 1913 of two Roman Catholic missionaries on the Barren Grounds, and the killing of the trader Robert Janes in 1920 on Baffin Island, were carried out according to Inuit custom. In both cases the victims had been behaving so erratically that they were considered to be dangerous, and their executions had been decided by the communities. Consternation and worry greeted the arrival of police, the arrest of the appointed executioners, their trials and their sentences to captivity in southern prisons. These were the first of many lessons in the duties of citizenship in a large nation, in which power and custom emanated from the temperate latitudes.

To the Saami of northern Scandinavia, who had been taxed for centuries by various governments and had experienced indentured labour in early mining projects, the late nineteenth and early twentieth centuries saw merely an acceleration in the long decline of their sovereignty. Faced with the population increases that fostered Scandinavian immigration to America at the time, and encouraged by the warming climates that followed the Little Ice Age, governments promoted the northward expansion of agriculture across the

Scandinavian peninsula. Pioneer farmers invaded the reindeer-grazing lands of the Saami, clearing forests that were needed by reindeer for winter browsing, and shooting reindeer that trampled their fields, considering them to be wild animals and free sources of meat. The Norwegian and Swedish governments passed laws designed to regulate the situation, but government policies generally saw the Saami as a hindrance to northern development and as people who had best be assimilated into the general populations of the Scandinavian countries. The Saami in the first half of the twentieth century underwent the same experiences as their neighbours in the Arctic regions of North America and the Soviet Union: forcible settlement in communities, and their children removed to residential schools that forbade the use of their native language.

The other country of the Nordic Arctic—Greenland—saw the extension of Danish law throughout the island, but this was accompanied by paternalistic economic and social policies directed to maintaining the local culture and shielding the Greenlandic population from exposure to the perceived dangers of the outside world. Commerce was regulated by a government trade monopoly, and fortunately for the people there were no major mineral or hydrocarbon discoveries. Like all other northern regions, however, Greenland was affected by the economic depression that swept the world during the 1930s. Southern markets for the fox furs, sealskins and other luxury goods produced by Arctic hunters suddenly collapsed; the halcyon days of the Arctic trapping industry were over, bringing to many northern peoples the realization that the way of life in which they were now enmeshed was at the mercy of decisions made in distant places, and of economic forces far beyond their control.

A new and unexpected phase of development came with the outbreak of the Second World War. Arctic outposts suddenly became strategic locations for prosecuting war, or for transporting the materials needed for combat. Questions of sovereignty and the ownership of Arctic lands, which had for generations been the subject of tedious paper-squabbles between southern governments, were now vital to national interests. Alaska became a potential Japanese invasion route to North America, as well as an air-bridge for ferrying aircraft from

America to their Soviet allies. With the Baltic closed to shipping, the Arctic ports of Murmansk and Arkhangelsk became essential to Soviet trade. The North Atlantic convoys carrying supplies from North America to Britain and the Soviet Union, as well as the German submarines that preyed on them, needed bases for communication, weather stations and harbours to which ships could retreat for repairs. Aircraft flying the North Atlantic required refuelling stops, and large military airbases were established at Goose Bay and Frobisher Bay in northern Canada, Narsarssuaq in Greenland and Keflavik in Iceland.

The sudden flood of military men accompanied by vast quantities of equipment and supplies transformed life in the regions of these bases, despite government policies that tried to isolate them from local populations. Most of these airbases continue to serve as transportation centres, and are still central to the communities that developed around them in wartime. Even in the most isolated locations, the remains of activities related to the Second World War are still apparent. In the Norwegian weather station on the tiny Barents Sea island of Høpen, I was shown a functioning radio transmitter inscribed with the German-language equivalent of the warning "Loose lips sink ships," a relic of the German occupation of this distant outpost forty years before. Another summer, while my friend Jim Tuck and I were searching for archaeological sites in northern Labrador, we stumbled on an untidy tangle of old wires, broken batteries and odd pieces of wood and rusted metal. We were mystified, but not knowledgeable or curious enough to try to solve the puzzle. A decade later a researcher working in German naval archives identified the debris as the remains of an automatic weather station, established in 1943 by a German submarine on this isolated stretch of the North American coast.

The transportation and communication technologies that were developed during the Second World War vastly increased the range of economic enterprise, administrative control and military presence across the northern reaches of both continents. By the end of the war, radio was capable of maintaining reliable communication links between the most remote outposts. This simple fact revolutionized

The radar screens and dome mark a station of North America's Distant Early Warning
system, established during the Cold War to warn of intercontinental bombers and missiles.
(Fred Bruemmer)

the efficiency and reliability of northern transportation; ships or
planes could now be called when needed, advised of local require-
ments and warned of local weather, sea-ice and landing-strip
conditions. It provided immeasurably improved security for the residents of isolated communities, who could now call for help in case
of accidents, food shortages or health emergencies. On both northern continents new bush-planes were designed to carry people and
freight into remote regions where they could operate from short
bumpy airstrips, snowy fields, ice or open water. The Soviet Antonov
AN-2, the most successful biplane ever built, first flew in 1947; the
first Canadian deHavilland Beaver took off in 1948 and the larger
Otter in 1951. All were remarkably similar, the design of their
engines and airframes grounded in the aviation advances of the
Second World War. For the next half-century, people living in isolated villages and outpost camps across the Arctic world listened for
the unmistakable rumble of distant radial engines, a welcome sound
that promised supplies, mail, a doctor or a ride home. They can still

be heard among the buzzing turboprops and whining jet engines that carry people and supplies to most Arctic communities today.

The invention and spread of nuclear weapons and the onset of the Cold War brought a new prominence to Arctic regions. For four decades after 1950 the circumpolar north became a frozen no-man's-land between two opposing nuclear-armed empires. Strings of radar stations mushroomed as nations rushed to provide themselves with forewarning of trans-polar nuclear bombers, and later of intercontinental missiles. Enigmatic outposts appeared on remote islands and icefields, providing weather information for the military aircraft patrolling Arctic skies, or monitoring the nuclear submarines that cruised beneath the polar pack. Small airstrips built to service radar and weather stations, as well as those in strategic locations that were designed to support the largest military aircraft, brought a new level of accessibility to most northern regions.

The military importance of the north focused the attention of southern governments on their Arctic territories as never before, and the latter half of the twentieth century saw a massive increase in the economic and social development of the north. Aerial surveys produced the first accurate maps of most Arctic regions, including the boundaries established by competing claims of sovereignty over polar lands and seas. Scientific research became a proxy for military occupation in demonstrating sovereignty over Arctic lands. Airborne scientists counted the wildlife, plotted the movement of sea ice and searched for mineral deposits. Newly developed prospecting techniques, together with the increased accessibility of the north, soon resulted in a broad scatter of mining ventures, and massive oil and gas deposits were discovered in the northwestern regions of both Eurasia and North America. On both continents, large tracked vehicles cleared winter-roads deep into the frozen tundra, hauling sled-trains of supplies for the establishment of military posts, mines and other commercial enterprises. Permanent roads and railways snaked northwards towards the new projects, linking previously isolated settlements to southern transportation networks and opening entire districts to tourists, sport-hunters and other southerners. Plans began to develop for the damming of northern rivers to supply hydroelectric power to the new northern industrial projects, or for exporting electricity to markets in the south.

The invasions of the 1950s and 1960s arrived at a time when imported diseases such as tuberculosis were still ravaging most Arctic communities, and when the fur-trapping economy in which many northern hunting peoples had become inextricably enmeshed had not yet recovered from the collapse of the 1930s. To Arctic peoples the accelerating incursion of southerners brought the full and final realization that their lands, and the ways of life that were so closely tied to those lands, were no longer their own. Most drastically affected were those who were removed from their homelands, usually against their wishes and for reasons beyond their comprehension. The reasons varied—several communities in the Soviet Union were removed from sensitive military zones, while some groups in Canada and Greenland were relocated in order to stake a claim to national sovereignty over territory that had previously been uninhabited. The flooding of hydroelectric reservoirs, the destruction of tundra by the toxic effluent of heavy industry and the contamination of fish resources all required the removal of local communities. As schooling and medical services were extended to northern populations, economic forces everywhere forced the abandonment of small villages and the resettlement of their residents into the larger and more efficiently serviced communities in which most northerners live today.

In every Arctic country these resettlement decisions were made by southerners who usually saw little consequence in moving people from one anonymous patch of tundra to another, from one icy river to the next, from a tiny coastal community to the larger town in the next bay. But for all traditional Arctic peoples, whether hunters, fishers or reindeer-herders, an intimate knowledge of their local environment is the key to survival. People identify themselves with the place they know, and with the other families who are identified with that place. Whatever the reason for resettlement, it was always a wrenching separation, tearing people from a land that had been part of themselves. A new environment, which to a southern administrator might look identical to the old, could be as disorienting as the relocation of an urban southerner from Toronto to Tokyo. Those

resettled usually found themselves among strangers who had no reason to welcome their arrival, and many of the tensions that strain
Arctic communities began in this forced loss of homeland at the
hands of administrative decision-makers.

The outright loss of land through resettlement is the extreme version of a disruption experienced by all northerners over the past
half-century. Even those who continued to live and work on the lands
of their ancestors found their traditional ways of life overturned or
made impracticable by the influx of southerners chasing well-paying
jobs and other opportunities in the north. In the more economically
developed regions of northern Russia, Alaska and Scandinavia, indigenous people quickly found themselves marginalized as a small and
easily ignored minority among the new northerners. Sportsmen
claimed their right to take the animals of what they assumed was
vacant wilderness, the same animals that had always been identified
with the lives of the indigenous groups whose lands they shared. Local
people were the last to learn of developments that might remove their
traditional livelihoods. They knew that when it came time to decide
whether to flood a local valley, to build an oil pipeline, a smelter or a
harbour, or to make arrangements for the sale of local resources and
the employment of local people, the decision would not be theirs.
Indigenous people were everywhere viewed as impediments to development, not as the owners of land and resources or as people who
might have an interest in these resources.

The loss of control over their land and their lives pushed the
indigenous peoples of the circumpolar world into a new era of political awareness. In most regions this awareness was forced by a major
industrial development that would affect the land that they occupied
or their use of the land. The 1968 discovery of major oil and gas
deposits in northern Alaska, the consequent Canadian plans to build
an oil pipeline down the Mackenzie Valley, the 1971 announcement
of an immense complex of hydroelectric developments in the James
Bay region of northern Québec and the Norwegian Government's
1978 scheme for a similar development in the heart of the Saami
homeland are prime examples of this process. Each stimulated opposition from the indigenous occupants of the lands that would be

affected by the development. To the younger people who had been educated in the languages and practices of the dominant society, the civil rights movements and environmentalist movements of the 1960s and 1970s provided models for effective resistance. Small political organizations began to form to oppose the decisions made in distant places, to influence those decisions or at the least to obtain some benefit from those decisions.

As these organizations grew and learned to speak effectively for the wishes of the original peoples of the Arctic, they began to address questions beyond the immediate damming of a river or the building of a pipeline. Everywhere the same concerns were spoken of: the need to preserve a culture and a way of life, the need to protect the local environment from destructive uses, the need to build a sustainable economy for the generations to come. And everywhere people gradually realized that these concerns were all intimately tied to ownership and control of the land itself. The idea that land can be owned—by a person, a community or a nation—is foreign to the thinking of most northern peoples, but they have learned that not to claim ownership leaves the land vulnerable to those who would use it in destructive ways. The view of the land as a being, as something large and strong yet requiring the protection of those with whom it is identified, is eloquently expressed in the charter adopted in 2001 by the central organization of indigenous peoples of the Russian north:

> Our home is the tundra, the taiga, the steppe and the mountains bequeathed to us by our ancestors; these are great, powerful, harsh, kind and generous manifestations, but they are defenceless in the face of technical progress.

The political associations that developed among the people of northern Russia after the fall of the Soviet Union express the same concerns as those representing the natives of other Arctic regions. And in Russia the reaction is not only to decades of exploitation of their lands by outsiders, but to the collapse of the system that enmeshed them in the economic and political life of the larger society. The past decade has been one of increasing hardship for the people of the Russian north. All of Russian society was affected by the economic implosion that followed the breakdown of the centrally planned economy, the problems of privatizing enterprises that had previously been the property of the state, and the shock of adjusting personal motives and community attitudes to the goals of capitalism. The new regime resulted in a major readjustment in the distribution of wealth, concentrating it in Moscow and other urban areas. While economic life has been difficult for many urban Russians, their distress is minor compared to that of rural populations, and the indigenous peoples of the most distant regions are those who have felt the harshest effects of post-Soviet life.

Many of the marginal industries that had been developed to employ northern people—fox-farming, commercial hunting and trapping, fish-processing—proved to be unprofitable once heavy government subsidies were removed. In some regions reindeer-herding collapsed along with subsidized markets, leaving state farms that could no longer pay wages to their workers, supply veterinary care to the animals or provide fuel and spare parts for herders' transportation. Many of the ethnic Russians who had lived in the north and collected high Soviet wages returned south, taking with them the technical and professional skills that had provided services to northern peoples. From the Kola Peninsula to Chukotka, the peoples to whom the north is home have been forced to improvise new ways of making a living. They pick mushrooms and berries, catch and preserve fish, compete with military poachers and commercial hunters for the game animals of the tundra and taiga, and contend for reindeer pasture with the booming oil, mining and forestry industries that produce most of Russia's export commodities. They miss the security and relative comfort that they enjoyed under the Soviet system, when they were contributing members of a huge, comprehensive social order instead of the disregarded remnants of a social experiment that failed.

The new Russian economy, open to global markets and global investment, has brought other difficulties to the peoples of the north. Far from ending the haphazard economic and industrial planning of the past, the new era has produced its own legacy of unrestricted and unregulated development. The oil and gas fields discovered in the lands of the Nenets during the 1980s have been a bonanza for newly privatized Russian companies, as well as for foreign firms that have purchased leases to drill wells and pump gas and oil to the insatiable markets of Europe and America. In the same way that the indigenous peoples of Alaska and northern Canada reacted to the developments of the 1970s, those whose lands are vulnerable to flaring gas and leaking pipelines feel themselves to be outsiders whose concerns are of little consequence to the hydrocarbon industry and the governments with which they work. Private ownership of Russian resources has produced situations that would be unthinkable in the West: the Ponoy River, the largest salmon river in the Kola Peninsula, is currently leased to an American sports-fishing outfitter and off-limits to the Saami for whom its runs of salmon have always been a crucial food resource.

As noted earlier, it was the threat of such developments that stimulated the formation of political organizations among the peoples of the Arctic world outside the Soviet Union during the 1960s and 1970s. The negotiations that these groups held with the various governments that claimed sovereignty over their homelands eventually led to a diverse range of accommodations, returning to the peoples of the north some measure of control over their lands and the ways of life that these lands supported. The most comprehensive shift of power occurred in Greenland, where the desire for political independence developed by a well-educated native Greenlandic intelligentsia coincided with the wishes of the Danish people to end their complicity in a colonial relationship. This agreement produced the Greenlandic concept of Home Rule, through which political control over most aspects of economic and social life was transferred in 1979 from Copenhagen to the new Greenlandic capital of Nuuk. Denmark retains the administration of foreign and defence matters, together with a financial commitment to the development of its erstwhile colony, but Greenland is essentially a distinct nation with its own independent government.

The American government took another route to accommodate the concerns of the indigenous peoples of Alaska, as well as those of

Caribou graze on the tundra near an oil rig at Prudhoe Bay, North Slope, Alaska. The 1971 Alaska Native Claims Settlement Act opened the way for major petroleum developments in northern Alaska. (Bryan and Cherry Alexander)

non-indigenous peoples who lived in the state or who wished to develop the lands that had originally belonged to the natives. The claims of indigenous people to original ownership of the land were extinguished by the Alaska Native Claims Settlement Act passed in 1971, in return for which native communities were paid almost a billion dollars and given title to parcels of Alaskan land totalling almost 200,000 square kilometres. The land and the money were signed over to twelve regional corporations that were established to administer the land claim, and to manage these resources in the interests of local indigenous groups. This approach was based on the premise that the resources paid in exchange for title to Alaska would allow the descendants of the original title holders to join with the non-native citizens of the state in developing a prosperous and cooperative future for all.

The concept of land claims settlement was also adopted by the Canadian government in dealing with the peoples of the north, but the largest and most notable of these settlement agreements also contains elements of the Greenlandic solution. The Nunavut Land Claim was settled in 1993, transferring 350,000 square kilometres of Arctic Canada, together with a large fund of money, to a corporation representing the Inuit of the central and eastern Arctic. At the same time, a process was undertaken that resulted in the 1999 establishment of Nunavut Territory, a largely self-governing region covering two million square kilometres (20 percent of the area of Canada) and coinciding with the area covered by the scattered Inuit land-holdings of the Nunavut Land Claim. With a population that is overwhelmingly Inuit, Nunavut (the name means "Our Land" in Inuktitut) is administered by an elected government bearing responsibility for most aspects of the economic and social life of the region. Like their Greenlandic neighbours, the Inuit of Nunavut have recovered a large measure of sovereignty over the lands originally occupied by their ancestors.

Yet another approach to accommodating the claims of indigenous northern peoples has been followed by the governments of the Scandinavian countries. The long-standing attempts to assimilate the Saami populations of northern Scandinavia into the dominant societies of Norway, Sweden and Finland began to be questioned during the 1960s, and by the 1980s had been officially abandoned. In response to fierce Saami opposition to their 1978 plans to dam the Alto-Kautokeino River, the Norwegian government led the way in developing a new policy that gave its Saami citizens some control over the regions of the country that they occupied. By 1989 this had developed into the Saamediggi, a parliament elected by the Saami and governing many aspects of Saami economic and social life. Sweden formed a similar body in 1993, Finland in 1996 and by 2000 a Saami Parliamentary Council was formed with representatives from all three countries and observer status for the organization representing the Saami of neighboring Russia. The relations between these ethnically based assemblies and the national governments that formed them are in a constant state of negotiation, with a focus on co-management of the resources of lands identified with the Saami, the preservation of Saami language and culture, and the development of a sustainable economic future.

To suspicions that the Saami may have ambitions to establish a trans-national country stretching across the north from the Atlantic

Ocean to the White Sea, the Saami reply that they have always dealt with too many national boundaries and do not wish to create even more. The same view is espoused by the Inuit Circumpolar Conference, an organization that represents the interests of Greenlanders, Canadian Inuit and the Eskimo peoples of Alaska and Chukotka. Like the Inuit, the Saami are more interested in ownership of the land with which they are identified, land that currently is owned primarily by state governments and used to benefit all of the citizens of the Nordic countries. Sven-Roald Nystø, President of the Norwegian Saamediggi and of the Saami Parliamentary Council, concluded a speech given in 2002 with the statement "The land must be returned to the people." When I asked him how this would be accomplished, he admitted that the Saami were much farther from this objective than were the indigenous peoples of Canada or Alaska, and that anything like a land claims settlement was impossible under current political conditions. Yet the Saami, like other northerners, have concluded that ownership of their own lands is central to the attainment of their goals of cultural and economic survival.

In northern Russia the question of land ownership is complicated by the fact that for seventy years the concept of private ownership of land was anathema to the communist society in which all but the youngest Russians have lived their lives and received their education. Supposedly drawing their inspiration from traditional Russian peasant practices, Soviet officialdom shared with indigenous peoples the attitude that private possession of such a basic commodity was incomprehensible. Such an attitude is not easily routed by the forces of market capitalism. Since the collapse of the state farms and other collective enterprises that had organized the Soviet Arctic economy, the Russian government has issued several decrees designed to ease the transition of indigenous populations to the new conditions, and to provide them with distinctive political rights. Although various forms of ethnically based regional governments have been proposed, the actual transfer of land to individuals or communities has not been undertaken. The most common approach has been the creation of *obshchiny*, family or clan-based groups that are formed to engage in a cooperative enterprise of reindeer-herding, fishing, sea-mammal hunting or other traditional land-use practices. Such groups can be formed only by indigenous people, and are granted leaseholds to areas of land which—in practice—frequently coincide with the lands that the family or clan occupied before or during the period of Soviet collectivization. Although *obshchiny* pay neither rents nor taxes on their landholdings, their proprietary interest is limited to the use of surface rights and falls short of what is generally considered outright land ownership in Western societies. On the political level, the legacy of communal organization left by the Soviet system has encouraged the development of a large number of overlapping committees based on local and ethnic interests, and an effective national organization capable of representing their rights and interests to the national government and to international organizations including the United Nations.

The political developments of the past few decades have provided most northern peoples with some measure of security against the distant and powerful interests that control the lands that have always been the basis of their lives. The effects of land claims, legislative representation and modifications to land tenure systems have been to return indigenous land use rights to some approximation of their condition a century ago. Many families or communities have at least the possibility of gaining stewardship over specific pieces of land, and their recognition by society as the people who are identified with that land is now at something like the same level that was enjoyed by their ancestors. However, all northern peoples now live in societies that are radically different from those of a century ago, and none wish to return to the life of those earlier societies. Like their ancestors who greeted the first European traders, whalers and missionaries, northern peoples expect to be full participants in the modern world. The political developments of the past decades provide only a background to this participation, much in the way that those of us from more densely occupied regions take for granted the background of rights and obligations administered by the governments of the municipalities and provinces or states in which we live. More important in our daily lives are the relationships with our families, friends, co-workers and others whose daily lives are bound to our own through complex webs of emotions and obligations and mutual support. But on a very different level, and perhaps as important in determining how we view ourselves, is our relationship to the culture and society of the larger world.

For most northern peoples, this larger world has been represented primarily by the growing stream of traders, missionaries, police officers, administrators, teachers, military personnel and others who have formed an increasingly important presence in Arctic communities over the past century. Most of these individuals arrive with an authority that derives from the office that they hold or from the economic benefits that they can distribute. But perhaps as important is the authority that derives from the expertise they are assumed to possess on the relationship between their own society and that of the northern community in which they are visitors.

Southerners move to Arctic communities for a great variety of reasons, and bring with them an equally wide range of knowledge and opinion regarding the people among whom they find themselves. Some have equipped themselves with information on the Arctic and its people, and if they arrive with stereotyped ideas, they are prepared to change them when they recognize that the reality is different; these people may spend their time in the north accumulating an unmatched knowledge of the region. However, these southerners must be balanced against the larger number who have little interest in accumulating first-hand knowledge, who are unwilling to let go of their preconceived ideas about Arctic communities, and who perpetuate the view so prevalent in the south: that the indigenous peoples of the region are the products of a primitive society that has not prepared them for full participation in the modern world.

This has been the view underlying the efforts made over the past century by national governments to educate indigenous Arctic people, to "prepare them for the modern world," and to assimilate them wherever possible to the languages and cultures of the south. It was the belief of most of the missionaries who laboured to save their primitive charges from the greater world, or to prepare them for exposure to that world. It is the perspective of the tourism industry that exhorts its customers to take an exotic vacation among the cheerfully primitive peoples of the north before they and their ancient ways of life disappear beneath the onslaught of globalization.

A generation ago, when I first visited the Arctic, the view that Arctic peoples were primitives unprepared for modern life was frankly expressed by most southerners living in the communities of Arctic Canada. It was an opinion I heard over and over, in Russia and Alaska, in Finland and Norway. One hears it less frequently these days, and perhaps we are learning. In some communities it may not even be the majority opinion among resident southerners.

Yet the idea that Arctic people are an archaic human type, the unchanged remnant of an ancient society, has an undeniable power. Even those who have acquired a deep knowledge of human societies can fall prey to its fascination. In his fine book *The Other Side of Eden,* anthropologist Hugh Brody relates that before his first visit to Arctic Canada he asked himself "... was I embarking on some kind of time travel? Was I leaving my own era and making a journey into the world as it was thousands of years ago?" Yet he found that "When I first got to know Anaviapik in Pond Inlet, he did not present himself as a human being from some other, more remote time." The Arctic of exploration accounts, of missionary literature and of the travel writers who have provided us all with images of a distant land frozen in time, the imaginary Arctic, is a very resilient fantasy even when confronted directly with reality.

Arctic people are well aware of the pervasive view that they are simple savages unprepared for modern life. Racial tolerance has never been a quality of Arctic cultures, and the contempt expressed by southerners has generally been returned in full. This response was sufficient to preserve people's confidence in themselves and their societies at a time when they had to deal with only a few individual southerners, but it is less effective when the extent of southern influence becomes overwhelming. In economically developed regions of Alaska, Russia and Scandinavia, where natives form a minority population in what were originally their homelands, the culture and the opinions of the larger world pervade all aspects of life. Even in remoter regions of the Arctic, where indigenous people still form a strong majority population, satellite television and other media flood people's consciousness with images of the greater world "outside."

Like rural people everywhere, those in Arctic communities are fascinated by the excitement, affluence and cultural wealth of the city. Even those who have no desire to visit or move to the city—be it Moscow or Copenhagen, Anchorage or Montreal—inevitably compare their lives with those described by displaced southerners, or portrayed on the screens of television and the internet, and the social and psychological troubles that are so apparent in Arctic communities must relate at least in part to the contrast that is perceived. A constant challenge to one's way of life by those who claim to represent a larger, richer and superior society must inevitably take a toll.

The past few decades have seen an increasing willingness to accept northerners as peoples of the modern world. In part, this sense has coincided with the development of political power by the indigenous peoples of the north, and the demonstration that they are capable of managing their lives without the constant guidance of experts from the south. But it is also the product of a gradual shift in the perspective of the larger societies that produced the experts, those who have created the image of the Arctic in the public mind. In *Canada and the Idea of North* Sherrill Grace notes a transformation of the way that the north is portrayed during the latter half of the twentieth century. At the beginning of this book I noted her description of the books and films of the earlier century, works that picture the north "... as a space for virile, white male adventure in a harsh but magnificent, unspoiled landscape waiting to be discovered, charted, painted, and photographed *as if for the first time*." She notes that after about 1950 this view is slowly replaced by one in which Arctic natives are represented not as exotic beings but as fellow humans: "This narrative development is crucial... for its representation of North as lived in, as occupied by people with stories and voices of their own." Then, in the work of late twentieth-century writers "northern space is often represented as occupied, as full of histories, stories, myths, and voices. The narrative that begins to emerge from these texts is hybrid, heterogeneous and unstable; the historical record, where it is evoked, is fragmented, questioned, rescripted." The North is no longer that of the southern visitor who travels through the country and emerges as an expert capable of informing the rest of the world of the truth.

Although Sherrill Grace deals primarily with the literary transformation of the Canadian north, the same process is apparent elsewhere. The prime example is perhaps Peter Høeg's 1993 novel *Smilla's Sense of Snow*, written in Danish and translated widely as well as produced as an English-language film. The heroine, Smilla Jasperson, is not a southern traveller but a Greenlander who combines scientific expertise related to ice with a deep and intangible feeling for the substance and for the icy land of her birth and ancestry.

Publication in the Greenlandic language has a history of well over a century, and has produced written and artistic works that are occasionally translated into Danish and other languages. A Saami-language publishing industry has flourished in the past thirty years, and again some of these publications appear in Scandinavian languages. The Soviet system supported limited educational publishing in indigenous languages, but the main outlet to the world at present—and a surprisingly effective one—is through the numerous internet Web sites developed by indigenous organizations. Alaskan and Canadian land claims agencies concerned with education and cultural survival support the publication of traditional narratives and memoirs of life in the old days, as well as at least one novel published in the Yupik Eskimo language.

Film has probably been the most effective medium in making the world aware of the views of Arctic aboriginal artists. The first to reach broad international distribution was the 1987 film *Ofelas (Pathfinder)* by Saami director Nils Gaup. Set in the ancient past and based on a Saami legend of murder and revenge, *Ofelas* was nominated for an Oscar as best foreign film of the year. An even wider distribution has recently been achieved by the 2001 film *Atanarjuat (The Fast Runner)*, the dramatic retelling of a Canadian Inuit legend by director Zacharias Kunuk, produced in Inuktituut with subtitles in world languages. The film won a major award at the 2001 Cannes Film Festival, and has given a global audience a rare view of the Arctic world through the eyes of an indigenous team of artists. Although the film reconstructs an ancient Inuit way of life, the characters are not remote and alien primitives, but individuals whose humour, anger and passion are as recognizably human as those of one's own family.

The reclaiming of the image of the Arctic by aboriginal people is perhaps as important as has been their reclaiming of traditional ownership and use of the land. No longer is the Arctic of the imagination the preserve of people whose minds and experience are formed elsewhere, and who see the Arctic as remote and dangerous and wild. The Arctic is in process of transformation from a land of the imagination to a place in the real and everyday world, a place that is inhabited by people no more bizarre and no less competent than those of more temperate zones.

This clear-sighted view of the Arctic will be needed if the peoples to whom it is home, and the national governments that claim sovereignty in its territories, are to cope with the problems that are converging on the region from several directions. The crisis that currently receives most attention is the accumulating evidence that world climates are warming, and that this process can be expected to be most pronounced in the Arctic—the very region for which cold has been the defining attribute. A warming climate will affect the Arctic most directly through the medium of sea ice. Ice that is thinner than normal, and that breaks up earlier in the summer, will drastically affect the populations and distributions of the animals that use the ice as protection, as a platform from which to hunt and travel, and as a feeding-zone. Already the polar bear populations of Hudson Bay are suffering from the seasonal absence of the ice that they need for hunting and travel. Indigenous hunters here and elsewhere use the ice in much the same way as bears, and are also experiencing the effects of unexpected changes in what had been a dependable element of their environment.

Sea ice is not the only element that will be affected by climatic warming. In the summer of 1996 I was confronted with the hideous consequences of a single warm weather event of the kind that is occurring with increasing frequency in recent years. For a week that summer I flew by helicopter over the beaches and bleak valleys of Bathurst Island in High Arctic Canada, making an inventory of archaeological sites on the island. I was sharing the helicopter with Frank Miller, a wildlife biologist who was continuing a survey of caribou and muskox populations. Of the hundreds of animals that had occupied the island the previous autumn we saw only 34 living muskoxen and 13 living caribou. Frank knew what had happened; in January, a month when the temperature in this far northern region normally goes no higher than -20°C and the rare precipitation falls as light snow, there had been a rainstorm. The muskoxen and the caribou shook off the rain falling in the midwinter night, but when the temperature dropped back to normal the snow that had turned to slush became an armour of ice that the animals could not dig through to reach the browse on which they depended for survival. Most of the caribou had disappeared, some perhaps crossing the ice to other islands, although the transmitters on caribou that Frank had radio-collared showed that the animals had not moved since February or March and their small bodies probably lay buried in snow. The frozen corpses of muskoxen were everywhere, mostly in family groups that had lived and starved together. Some still stood in snowdrifts where they had bogged down and died, too weak to continue their futile search for food.

Even if the warming conditions of the past decade prove to be a short-lived episode, rather than the forerunner of a major change in the earth's climatic pattern as many believe, its results will bring hardship to the animals and people who can no longer predict the environment on which they depend. The unpredictability of weather in recent years devalues the knowledge that is the most basic resource of Arctic hunters and travellers. This is a recurring theme of *The Earth Is Faster Now*, a recent book by Igor Krupnik and Dyanna Joly devoted to indigenous observations of climatic change in the Arctic. A Bering Sea Eskimo commentator reports an elder telling him that "the Earth is faster now.... She was not meaning that the time is moving fast these days or that events are going faster. But she was talking about how all this weather is changing. Back in the old days they could predict the weather by observing the stars, the sky, and other events. The old people think that back then they could predict the weather pattern for a few days in advance. Not any more!" The townsmen and commercial operators of the north will face different but equally serious problems. The most important will be the melting of the permafrost, the platform on which buildings, roads and airstrips are constructed—all the vital underpinnings of modern life in the Arctic.

At a very different level, the expected thinning of polar sea ice has raised expectations of new ice-free shipping routes across the northern edge of the continents or even over the pole. The Northeast and Northwest passages may at last become commercially viable realities in the coming decades; in fact, the first commercial use of the Northwest Passage across Arctic Canada was made during the warm summer of 1999, by a Russian icebreaker and tug towing a floating drydock from Vladivostok to its new owner in Bermuda. That a changing environment has now made such ventures possible has raised political anxiety regarding sovereignty and the policing of regions that until now have been protected by ice. Pollution concerns, unpoliced access to the resources of remote regions, the potential for military use of such areas, even the use of distant and once unreachable Arctic islands as bases for terrorism and illegal immigration, are now part of the Arctic agenda discussed by northern nations.

The close linkage of the Arctic climate to that of more southerly latitudes means that the Arctic is not isolated from the problems caused by industrial pollution in the lands to the south. Aside from the smelters of Norilsk and the Kola Peninsula, nuclear waste buried or sunk beneath the sea in a few scattered areas, and the gas-flaring and oil-spills associated with northern oilfields, the Arctic creates very little industrial pollution. However, the prevailing directions of winds, together with ocean currents and north-flowing rivers, bring the region more than its share of contaminants from other parts of the world. The most vivid example of this long-distant transport was the Chernobyl incident of April 1986, when an explosion at a nuclear generating plant in Ukraine spread a plume of radioactive particles northward across Eurasia. Although the effects on the health of most northern Europeans seem to have been relatively slight, the tragedy had a more dire consequence for Saami reindeer herders. Radioactive cesium became incorporated in the spring growth of tundra lichens and was concentrated in the flesh of the reindeer that fed on the lichens; more than 20,000 animals had to be slaughtered and discarded.

Although the Arctic environment is still almost pristine by southern standards, the Arctic's polar location and extreme seasonal changes in light and temperature, together with its unique biological processes, make the region particularly vulnerable to the contaminants that arrive from the south. The polar zones, for example, experience the formation of springtime "ozone holes" in the upper atmosphere, exposing the regions beneath them to potentially dangerous levels of ultraviolet light. Although the manufacture of chlorine compounds that cause this phenomenon has been slowed by international agreement, their effects—together with the compounds unpredictably released by volcanic eruptions—will continue to affect the polar regions for decades. The widespread sense among northerners that the sun has felt hotter in recent years may, in fact, owe something to this phenomenon.

The short Arctic spring is another cause of vulnerability. The rapid melting of snow and the sudden flush of contaminants that have accumulated all winter produces a chemical jolt that is more destructive than the usual effect of low-level contamination. Mercury reaching the atmosphere from burning coal, and organic contaminants associated with pesticide use in temperate or tropical regions, surge into the Arctic food-chain during the spring and are passed along and concentrated at each level until they reach the fish and sea mammals, and eventually the bears and humans that prey on these creatures. The organic compounds that are of specific concern to health are particularly soluble in fat, and since Arctic animals are so dependent on fat as insulation and energy storage in the cold season, they are exceptionally vulnerable to these contaminants. As a result, some Inuit in Greenland and Arctic Canada have contamination levels that are among the highest of any peoples on earth, to the extent that breast-feeding of infants is now discouraged.

These levels of contamination are most pronounced in the people who are most dependent on a traditional hunting lifestyle. Individuals and families are thus forced to choose between the health risk of contaminated natural food and the risk of diabetes and other problems associated with the consumption of southern foods. Food imported from the south is expensive and, aside from sugar-based items, it is generally unappealing to those who have been raised on a diet of meat and other country foods. Giving up foods that have been a life-long source of sustenance and comfort is difficult for most people, and the sense that these foods are no longer clean and healthful is disheartening to those who continue to use them from choice, or because they cannot afford to do anything else.

This is only one of the dilemmas that the indigenous residents of northern communities are facing, but it is emblematic of the wrenching changes they are experiencing. The problems of a warming climate, ozone depletion and contaminated food sources are frightening, but for many individuals they are only low-level background concerns in comparison with the severe problems that arise from the social conditions of northern communities. When the traditionally high birth rates of Arctic peoples are combined with the effective medicine and economic security of the modern world, a rapidly increasing population is inevitable. Aside from northern Russia, where the current disastrous economic conditions have produced a falling birthrate and a lowered average age at death, the Arctic is experiencing population growth at a rate that outstrips most of the world. Northern towns are everywhere overcrowded, with housing and community facilities that would be judged clearly inadequate in the temperate regions of the European and American nations of which they are a part. In the schools, classrooms are crowded, teachers attempting to cope with cultural problems are disgruntled and frustrated, and students are diffident and unhappy. Funding is inadequate for the unique difficulties of northern education. Except in the few regions where the local economy is securely based on resource extraction, there are no jobs or other opportunities for most of the ever-growing stream of students graduating from the schools. Geographical and cultural isolation prevents these kids from escaping to the city or to a more affluent region of the country, as young people who live in economically deprived regions of the south traditionally do.

The economic difficulties facing most northern communities often force individuals to choose between a traditional land-based economy that does not supply everything needed for life in a modern town, and the education-based economy that may not be capable of providing any life at all except in unthinkable exile to a distant city. Mixed with these are the cultural and social choices faced by people from small minority communities dealing with the confident authority of world languages and cultures, of poor people confronted by the inexplicable wealth of others. The addiction problems that plague all northern communities, the suicide rates that far outstrip those of the south and the accidents that kill so many young men are too easily blamed on the inadequacies of primitive northern cultures. In fact, the peoples of the north are coping with a complex of problems that would stun most southern communities, and the symptoms of distress are balanced by a remarkable resilience, a sense of optimistic confidence and hope, a belief that whatever the future brings can be dealt with and surmounted.

Although the past century has seen a remarkable transformation in the lives of Arctic peoples, the difference from previous centuries has been only one of scale. Change has always been a part of Arctic life, as communities and cultures transformed themselves to accord with their changing surroundings. The indigenous peoples of the Arctic have coped with recent changes through the same means used by everyone else in the modern world: by educating themselves in the skills of the present, by attempting to reclaim political control over their lives and by gaining a growing measure of artistic influence over the way in which the world imagines them and their lands. The real legacy that northern peoples inherit, from ancestors who have lived through so many changes in the Arctic world, may be the confidence with which they greet the problems of the present as well as those that any future may bring.

13 AN ARCTIC JOURNEY

THIS BOOK HAS BEEN AN ATTEMPT TO understand our vision of the Arctic as a world apart, where past and present merge in a distant and compelling landscape of brilliant light. In tracing the gradual construction of this image we have journeyed through deep time, beginning in the Ice Age environment of our early human ancestors, which may yet resonate at some unfathomable level of cultural memory. We have visited the territory of the carefree and immortal Hyperboreans described by the geographers of the Classical Mediterranean world, and have glimpsed Ultima Thule protected by the congealed and heaving sea encountered by the first Arctic explorer of that time. We have followed these images through two millennia of speculation, exploration and creative artistry, and seen them layered with the hardships and hideous trials endured by the unprepared explorers of the past few centuries. We have witnessed how the image was shaded and enhanced by the conventions of Victorian travel literature, and by the novelists and artists and adventurers who have reported the Arctic to the twentieth-century world. We have looked at the recent stirrings of a movement among northerners who have begun to tell others about the Arctic world and the experiences of Arctic people. The products of these artists and writers and film-makers lack the quality of "otherness" that was central to the spirit of earlier descriptions of the polar regions. For the first time we have been given a view of the Arctic as a normal part of the world, a beloved homeland rather than a frightening alien environment. The Arctic is no longer an imaginary place where anything strange and terrible might happen.

In writing the book I found myself questioning the balance between experience and the enthralling aura of the unknown. Do knowledge and clear vision compensate for the loss of an imaginary world, or is it possible for the two levels of perception to coexist? Perhaps the southern vision of the Arctic is so enticing that it cannot be entirely submerged by reality. The Arctic may be so startling, so astonishing, that it overwhelms familiarity, continuing to shock the senses of the experienced southern visitor.

For myself, the enchanted Arctic of my first imaginings still breaks through the mundane world of familiar perceptions, most reliably on the rare summer nights when the wind has died, the silent landscape is bathed in a luminous golden glow from the almost-setting sun and sleep is impossible. Once from a midnight hilltop on Banks Island I listened to silence broken by the distant mew of gulls, the occasional grumble of drifting sea ice, and a loon drawing a line of sound across the empty sky; down the distant skyline the first muskox that I had ever seen travelled like a tiny mechanical toy on some lonely journey that might have begun during the Ice Age. On another such occasion I watched through the midnight hours, puzzled and wary, as a white bear sat like a dog among the flowers that fringed a shallow tundra pond near my tent; for two hours in the stillness she stared into the water, watching her reflection or perhaps imaginary seals, before wandering away to her next adventure.

And one night in the Inuit village of Qaussuituq I awoke, and from the window of my rented house saw the gravel street empty except for a tiny girl wandering purposefully in the golden light; in imitation of the babies carried by her older sisters, she had placed a stuffed toy animal in the hood of her parka. While the adults of the community tried to adjust their lives to the working timetables imported from the south, this lone child maintained the wakeful summer night that for her ancestors, and for all other Arctic peoples, has always been the time for hunting and travelling and experiencing the world at its magical best. At such times, even those from the remote south can share the exhilaration of distant uncluttered horizons, the freedom of endless light and the sense of being enclosed by a world of serenity. It is a sense of the Arctic world that no familiarity can dispel.

USEFUL REFERENCES,

INTERESTING READING

THE FOLLOWING DOES NOT PRETEND to be a comprehensive listing of all the sources used in the preparation of this book, nor of scholarly publications relating to the history of the Arctic regions. Rather, it contains a selection of books that I found both useful and interesting, most of which are available through bookstores and libraries. They are arranged according to the chapters of this book with which the references are most closely associated.

Prelude: An Arctic Vision

Canada and the Idea of North. Sherrill F. Grace (Montréal and Kingston. McGill-Queen's University Press, 2001). An analysis of the concept of "north" as interpreted in Canadian art and literature, and of how the concept has evolved over the past century. Grace describes with precision the process by which my fascination with the Arctic developed.

People of the Deer. Farley Mowat (Boston: Little, Brown, 1951); and *The Desperate People* (Boston: Little, Brown, 1959). Farley Mowat's classic tales of a small band of Inuit, vulnerable and helpless in their encounter with the modern world, were influential in establishing an image of the Arctic in the culture of the mid-twentieth century.

Between Heaven and Hell; The Myth of Siberia in Russian Culture. Edited by Galya Diment and Yuri Slezkine (New York: St. Martin's Press, 1993). A collection of essays tracing through the centuries the linked concepts of Siberia as both "the frightening heart of darkness and a fabulous land of plenty."

Echoing Silence; Essays on Arctic Narrative. Edited by John Moss (Ottawa: University of Ottawa Press, 1995). A wide-ranging series of articles by writers, historians, and native northerners, documenting the diverse ways in which the Arctic is observed and described both historically and in recent times.

Landscape and Memory. Simon Schama (New York: Alfred A. Knopf, 1995). An exploration of the links between mythology, history and geography, arguing that concepts such as landscape, scenery and wilderness are very much based in the historical associations of places and environments.

After the Ice Age
Ice Ages: Solving the Mystery. John Imbrie and Katherine Palmer Imbrie. (Cambridge, Mass.: Harvard University Press, 1986). A well-written book on the discovery of the Ice Age, and its contribution to the nineteenth-century clash of science and theology. The authors trace the development of explanations for glacial periods over the past 150 years, and portray current scientific thinking on the subject.

After the Ice Age. E.C. Pielou (Chicago: University of Chicago Press, 1992). The chapter heading echoes the title of this excellent book, which describes the process by which climates and environments developed from those of the glacial period to those of the present world. The emphasis is on North America, and humans are given little consideration, but the book provides an excellent background to the picture of ancient human history presented here.

A Distant Paradise

The diverse ideas discussed in this chapter are not easily referenced to a few accessible sources. However I will use this opportunity to introduce Jeannette Mirsky's *To the Arctic: The Story of Northern Exploration from Earliest Times to the Present* (Chicago: University of Chicago Press, Chicago, 1970). Originally published in 1934, this well-written and extremely readable book remains the most thorough summary of Arctic exploration yet published. Richard Vaughan's *The Arctic: A History* (Stroud: Alan Sutton, 1994) is a more current summary of the same material. Derek Hayes' recent *Historical Atlas of the Arctic* (Vancouver: Douglas & McIntyre, 2003) is an invaluable compendium of over 300 maps and details from maps relating to all phases of Arctic exploration.

Those readers who have not yet discovered the online resource known as Project Gutenberg, a massive international project dedicated to making available the texts of important and obscure works of historical and literary interest, can find this extremely useful source at http://promo.net/pg/index.html.

A Hunter's World

The Other Side of Eden: Hunters, Farmers and the Shaping of the World. Hugh Brody (New York: North Point Press, 2001). A study of the culture and psychology of hunting peoples, and how these determine the manner in which hunters relate to more complex societies.

Arctic Adaptations: Native Whalers and Reindeer Herders of Northern Eurasia. Igor Krupnik (Hanover and London: University Press of New England, 1993). English translation of Krupnik's important Russian-language book exploring and analysing the characteristics shared by Arctic peoples from the Ice Age to the present day.

Ancient People of the Arctic. Robert McGhee (Vancouver: UBC Press, 1996). A general-audience book describing what we know of the archaeology and culture of the Tuniit people.

In Arctic Siberia
A History of the Peoples of Siberia: Russia's North Asian Colony, 1581–1990. James Forsyth (Cambridge: Cambridge University Press, 1992). A comprehensive and scholarly study of the small nations of northern and eastern Russia, from the time of Russian conquest to the breakup of the Soviet Union.

Crossroads of Continents: Cultures of Siberia and Alaska. Edited by William W. Fitzhugh and Aron Crowell (Washington, DC: Smithsonian Institution Press, 1988). This heavily illustrated catalogue, published to accompany a major museum exhibit, is a collection of well-written and authoritative chapters on the history and traditional cultures of the peoples on either side of Bering Strait.

Arctic Mirrors: Russia and the Small Peoples of the North. Yuri Slezkine (Ithaca and London: Cornell University Press, 1994). An excellent treatment of the long and complex historical relationship between Russians and the aboriginal peoples of Siberia.

Siberian Survival: The Nenets and their Story. Andrei V. Golovnev and Gail Osherenko (Ithaca and London: Cornell University Press, 1999). A thorough examination of the history of the Nenets people of the Yamal Peninsula of northwestern Siberia. The authors trace the cultural survival of a small tundra nation from their first contact with Russian traders and tax collectors, through the Revolution, Civil War and collectivization of the Soviet era, to the breakdown of the Soviet system and the incursion of the oil and gas industry into their ancestral lands.

Vikings and Arctic Farmers
The Norse Atlantic Saga. Gwyn Jones (Oxford: Oxford University Press, 1986). The second edition of an older work, and still the best and most readable introduction to the adventures of the Norse in Iceland, Greenland and northeastern North America.

Vikings: The North Atlantic Saga. Edited by William W. Fitzhugh and Elizabeth Ward (Washington, DC: Smithsonian Institution, 2000). Catalogue for a museum exhibition mounted to celebrate the millennium of the Norse discovery of North America. This heavily illustrated collection of papers by a broad variety of scholars provides the most accessible and recent information available on the subject.

Inuit

Many of the ideas and much of the information presented in this chapter have accumulated over years of personal research on the subject. No good contemporary reference exists on the subject of Inuit history, but for a somewhat dated account the reader is best referred to Don E. Dumond's *The Eskimos and Aleuts* (London: Thames and Hudson, 1990).

Arctic: Handbook of American Indians, Vol. 5. Edited by David Damas (Washington, DC: Smithsonian Institution, 1984) is a compendium of anthropological knowledge relating to the history, society, culture and social development of the Eskimo and Aleut peoples, together with an excellent bibliography of sources published before about 1980.

The Eskimos by Ernest S. Burch Jr. (Norman and London: University of Oklahoma Press, 1988) is a comprehensive, authoritative and beautifully illustrated volume depicting both the adaptations and the cultural wealth of traditional Eskimo societies. The brief section on history is out of date, and the focus is on the Eskimo peoples of Alaska.

Inuit: Glimpses of an Arctic Past, by David Morrison and Georges-Hébert Germain (Hull, Québec: Canadian Museum of Civilization, 1995) is a similarly well-written and heavily illustrated book that focuses on the Inuit of the Canadian Arctic.

Inua: Spirit World of the Bering Sea Eskimo. William W. Fitzhugh and Susan A. Kaplan (Washington, DC: Smithsonian Institution Press, 1982). This heavily illustrated catalogue, compiled to accompany a museum exhibition, presents a comprehensive study of the rich Inuit societies of western Alaska.

Relocating Eden: The Image and Politics of Inuit Exile in the Canadian Arctic. Alan Rudolph Marcus (Hanover and London: University Press of New England, 1995) is an excellent summary of the resettlement of Inuit populations in Arctic Canada during the 1950s, the assumptions that led to the endeavours and the consequences that resulted.

Arctic Migrants/Arctic Villagers by David Damas (Montréal and Kingston: McGill-Queen's University Press, 2002) provides a more comprehensive documentation of the processes that resulted in the current Inuit communities of Arctic Canada.

Ice and Death on the Northeast Passage
The Three Voyages of William Barents to the Arctic Regions, 1594, 1595, and 1596, by Gerrit de Veer, Edited by Charles T. Beke, 2nd ed. (London: Hakluyt Society, 1876).

In the Land of White Death. Valerian Albanov (New York: The Modern Library, 2000). The first English translation of Albanov's astonishing narrative, this small book encapsulates the combination of folly, poor planning, luck and endurance that characterizes so much of European penetration of the Arctic regions.

Karluk: The Great Untold Story of Arctic Exploration. William Laird McKinley (London: Weidenfeld and Nicolson, 1976). A full account of the Karluk adventure, written in old age by one of the survivors, revealing remarkably little bitterness towards Vilhjalmur Stafansson whose actions caused the tragedy. McKinley's unpublished notes, from which he planned a more complete and accurate version, were a principal source for Jennifer Niven's excellent retelling of the story in *The Ice Master: the Doomed 1913 Voyage of the Karluk* (New York: Hyperion, 2000).

Martin Frobisher's Gold Mines

The Arctic Voyages of Martin Frobisher: An Elizabethan Adventure. Robert McGhee (Montréal and Kingston: McGill-Queen's University Press, 2001) provides a recent treatment of the subject, written for a general audience.

Martin Frobisher, Elizabethan Privateer. James McDermott (New Haven and London: Yale University Press, 2001). McDermott places the Arctic voyages in the context of Frobisher's life and times, and presents the only modern biography of the man.

The Three Voyages of Martin Frobisher. Edited by Vilhjalmur Stefansson and Eloise McCaskill (London: Argonaut Press, 1938). This compilation of documents contains many of the most important records relating to the Frobisher voyages, including the narratives of George Best, Dyonese Settle and Christopher Hall, as well as a detailed analysis of the venture by Arctic explorer Stefansson.

The Rape of Spitsbergen

No Man's Land: A History of Spitsbergen. William Martin Conway (Cambridge: Cambridge University Press, 1906). An old book, but still the best summary of the early history of Svalbard.

Seasons with the Sea Horses, or Sporting Adventures in the Northern Seas. Sir James Lamont (New York: Harper, 1861). A classic Victorian sporting adventure.

Walvisvaart in de Gouden Eeuw. Edited by Louwrens Hacquebord and Wim Vroom (Amsterdam: De Bataafsche Leeuw, 1988). A collection of reports and essays relating to Smeerenburg and the golden age of the Dutch whaling industry. This finely illustrated volume with English language summaries is useful even for those who, like the present author, do not read Dutch.

Bay of Tragedy

New American World: A Documentary History of North America to 1612. David Beers Quinn (New York: Arno, 1971). Volume 4 contains a selection of papers related to the explorations of Frobisher and Davis, as well as the surviving portion of Henry Hudson's journal, the narrative of Abacuk Pricket and court transcripts related to examination of the mutineers on Hudson's last voyage.

The Journal of Jens Munk 1619–1620. Edited by Walter A. Kenyon (Toronto: Royal Ontario Museum, 1980). This small volume republishes a scarce English translation of the journal describing the terrible events that occurred in Churchill River during the winter of 1620.

A Journey from Prince of Wales's Fort in Hudson's Bay to the Northern Ocean, 1769–1772. Samuel Hearne. Edited by Richard Glover (Toronto: Macmillan, 1958). Perhaps the most useful of several editions of this classic work.

Frozen Glory

I May Be Some Time: Ice and the English Imagination. Francis Spufford (New York: St. Martin's Press, 1997). A remarkable exploration of the English infatuation with polar exploration, drawing on literature, culture history and the works of the explorers themselves. A thoroughly enjoyable book that should be read as background to any study of Arctic exploration.

The Exploration of Northern Canada, 500–1920: a Chronology. Alan Cooke and Clive Holland (Toronto: The Arctic History Press, 1978). This massive compilation provides a chronological listing of basic information on hundreds of expeditions and events related to northern exploration, together with maps, indexes and an exhaustive bibliography. It is a fundamental reference for anyone interested in the history of Arctic North America.

The Search for the Northwest Passage. Ann Savours (New York: St. Martin's Press, 1999). The most recent, authoritative and detailed treatment of English ventures in search of a sea-passage to the north of North America, told from a viewpoint that is more admiring and uncritical than that of this chapter. The extensive and current notes make reference to most relevant sources, including the recently published journals kept by all the officers involved in Franklin's first overland expedition.

Qitdlarssuaq: l'histoire d'une migration polaire. Guy Mary-Rousselière. (Montréal: Université de Montréal, 1980). The most thorough investigation and narration of Qitdlaq's remarkable adventures.

Cook and Peary: The Polar Controversy Resolved. Robert M. Bryce (Mechanicsburg, PA: Stackpole Books, 1997). The most thorough treatment yet published of matters related to American polar exploration during the early twentieth century, and a remarkably even-handed analysis of the Cook–Peary debate. Despite the optimism of the title, the controversy regarding the attainment of the pole can be expected to continue.

The People's Land

The People's Land: Whites and the Eastern Arctic. Hugh Brody (Harmondsworth: Penguin, 1975). I owe the chapter title to that of this perceptive study of relations between the Inuit and outsiders living in their homeland, written at a time when the political developments outlined in this chapter were beginning to occur.

Peoples of the Tundra: Northern Siberians in the Post-Communist Transition. John P. Ziker (Prospect Heights, IL: Waveland Press, 2002). This recent ethnographic study of a community of hunters and reindeer herders in the Taimyr Peninsula provides an excellent description of the changes and stresses currently faced by the peoples of Arctic Russia. *Siberian Survival: The Nenets and Their Story*, noted earlier, also provides an excellent account of this situation.

The Earth Is Faster Now; Indigenous Observations of Arctic Environmental Change. Edited by Igor Krupnik and Dyanna Jolly (Fairbanks: Arctic Research Consortium of the United States, 2002). A collection of papers dealing with native perceptions of current environmental changes, and of the potential impacts of changes expected in the future.

For current information, the reader is best referred to internet sites maintained by some of the non-government organizations concerned with developments in the Arctic regions. Among these are:

http://www.inuit.org The Inuit Circumpolar Conference is an international organization representing approximately 150,000 Inuit and Eskimos living in the Arctic regions of Alaska, Canada, Greenland and Russia.

http://www.raipon.net/english/index.html The Russian Association of Indigenous Peoples of the North (RAIPON) represents approximately 200,000 people belonging to over thirty indigenous groups, from the Saami in the west to Eskimos in the east of the Russian Federation.

http://www.saamicouncil.net/ The Saami Council represents the interests of the Saami of Norway, Sweden, Finland and Russia to promote the interests of the Saami as a nation and to maintain the economic, social and cultural rights of the Saami in the legislation of the four states in which its members reside.

http://www.arctic-council.org/index.html The Arctic Council is an intergovernmental organization with members from eight circumpolar nations. It provides a forum to address the common concerns faced by the Arctic governments and the people of the Arctic, with particular interests in environmental matters and sustainable development.

An Arctic Journey

I have left to the last a strong recommendation for Barry Lopez's *Arctic Dreams: Imagination and Desire in a Northern Landscape* (New York: Scribner, 1986). Part informative travel narrative, part beautifully written personal meditation, this fine book has quickly become a classic of Arctic literature.

Index